THE I TATTI
RENAISSANCE LIBRARY

James Hankins, General Editor

ZABARELLA

ON METHODS

VOLUME I

ITRL 58

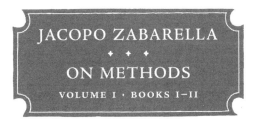

JACOPO ZABARELLA
✦ ✦ ✦
ON METHODS
VOLUME I · BOOKS I–II

EDITED AND TRANSLATED BY

JOHN P. McCASKEY

THE I TATTI RENAISSANCE LIBRARY
HARVARD UNIVERSITY PRESS
CAMBRIDGE, MASSACHUSETTS
LONDON, ENGLAND
2013

Series design by Dean Bornstein

Library of Congress Cataloging-in-Publication Data

Zabarella, Jacopo, 1533–1589.
[Opera logica. Selections. English]
On methods / Jacopo Zabarella ; edited and translated by John P. McCaskey.
volumes cm. — (The I Tatti Renaissance library ; 58–59)
Latin text with English translation.
Includes bibliographical references and index.
ISBN 978-0-674-72479-2 (v. 1) — ISBN 978-0-674-72480-8 (v. 2)
1. Logic — Early works to 1800. 2. Methodology — Early works to 1800.
I. Zabarella, Jacopo, 1533–1589. Opera logica. Selections. II. Title.
III. Title: De methodis. IV. Title: De regressu. V. Title: On regressus.
B785.Z23O613 2013
160 — dc23   2013017883

Contents

ॐ

Introduction

෧෧෧

Jacopo Zabarella spent his entire life in Padua, the university town of Venice, one of the greatest intellectual centers in the Europe of his day. He was born to a noble Paduan family on September 5, 1533. He received his doctorate in the arts from the university twenty years later. He was appointed to the first chair of logic in 1564 and—although he did not have a degree in medicine as was typical for the position—was in 1569 appointed to the junior position of extraordinary professor on natural philosophy in the second chair. He was promoted to the first extraordinary chair of natural philosophy in 1577 and the following year published a collection of commentaries, essays, and teaching tools under the title *Opera Logica*. His largest single work, a commentary on Aristotle's *Posterior Analytics*, was published in 1582. In 1585, he was promoted to second ordinary chair in natural philosophy, the highest chair a native Paduan was allowed to hold. At the time of his death at age fifty-six, on October 15, 1589, Zabarella was the most eminent writer in his field.[1]

He was soon forgotten. Zabarella's expertise was philosophy and science that took the writings of Aristotle as their starting point. Zabarella could read Aristotle in Greek. He knew Aristotle's logical works; Aristotle's cosmology, physics, and metaphysics; the biological works; even the minor tracts. He had access to commentators ancient, medieval, and recent. He knew his Aristotle well. But he also knew Galen, Avicenna, and Averroës, and he insisted that he was willing to criticize any of the ancients when he found them to be wrong. Still, he usually concluded that the ancient authorities were right and that it was the modern writers and teachers—especially the teachers of medicine—who were wrong. When he found disagreement among the ancients, Zaba-

rella invariably sided with Aristotle. Study of Aristotle had waned early in the sixteenth century but had revived by the 1540s. However, in the decades after his death, Zabarella's kind of Aristotelianism fell out of favor. Indeed, it represented much of what early modern philosophers and scientists were railing against.

Later historians of science and philosophy shared this disdain and gave Zabarella little attention. But at the beginning of the twentieth century, that changed. Historians including Ernst Cassirer and Pierre Duhem stressed that Gilbert, Galileo, Bacon, Harvey, and Descartes did not spring into being *ex nihilo*. To fully understand them, we need to study what they were reading and learning in their youth. Attention turned to late sixteenth-century discussions of method, to the University of Padua, and to Zabarella — especially after publication of John Herman Randall's 1940 essay "The Development of Scientific Method in the School of Padua."[2] Randall presented Zabarella and his method of regressus as the very seed from which the Scientific Revolution grew. Much English-language scholarship on Zabarella has been influenced — often overly influenced — by Randall's thesis.

Unfortunately, that scholarship has been hindered by a lack of readily available texts. The two volumes presented here provide the first modern Latin edition and the first English translation of the works that have garnered the most attention, *On Methods* and *On Regressus*.

ON METHODS

Zabarella begins *On Methods* with these words: "It is manifest that every science, every art, and every discipline is conveyed by some method." His verb is *tradere*, and it means "to deliver, hand down, pass along, convey." The noun is *traditio*, and in Zabarella's day, the words, in both Latin and English forms, were associated with the passing on of knowledge in the schools. In 1605 Francis Bacon

wrote that he gives "the general name of Tradition or Delivery" to "expressing or transferring our Knowledge to others." The first part of tradition, he says, is the "Organ of Tradition" and is either speech or writing. The second part is the "Method of Tradition" and, he reports, "it has moved a Controversy in our time."[3] The controversy had by then been going on for several decades. In England it had fueled a spirited debate at Cambridge between Everard Digby and William Temple. In western Europe, the controversy was dominated by proposals from the archcritic of conventional Aristotelian pedagogy, Peter Ramus. And in Padua, the controversy engaged innovators and conservatives in natural philosophy and in medicine. *On Methods* was a contribution to the debate from the leading logician at Europe's leading center of secular Aristotelian scholarship. *On Methods* is not a treatise about scientific discovery; it is Jacopo Zabarella's proposal for the proper way to convey knowledge from master to pupil.

Book I

In Book I, Zabarella distinguishes method and order and claims that, by its very definition, the order of teaching must always be from universal to particulars.

Early in the book, Zabarella explains that order and method are both instruments with which we come to know. Order — just as the English word means for us today — is the sequence in which the components of some treatise or treatment are presented. Method — in a usage less familiar nowadays — is how each step in the presentation leads to the next. "Order makes no inference of this thing from that, but only disposes the things to be treated. . . . Method, however, . . . leads us from the known into knowledge of the unknown, inferring the latter from the former" (*Meth. l.* 1 *c.* 3 *par.* 2). To teach correctly, one must use the correct order.

At the beginning of the *Physics*, Aristotle says that in any discipline that involves principles, causes, or elements, we must pro-

ceed in the natural order, from universals to particulars, from what is more known to what is less known. Averroës, Zabarella reports, said we must proceed from universals to particulars *because* universals are more known to us than particulars. Others say this cannot be right, since universals are so often less known to us, not more known. These others say that what is most important is to follow the natural order. Zabarella presents several arguments against this interpretation and bolsters his position by appeals to Aristotle's own practice in the *Metaphysics*, *Nicomachean Ethics*, and other works. Then, in two crucial chapters, 8 and 9, Zabarella presents the core of his own theory of order.

Several distinctions are central to Zabarella's theory. The first is the aforementioned distinction between order and method, including between the utility of each. Second is the distinction between imperfect and perfect knowledge. Imperfect knowledge is confused; perfect knowledge is distinct. Knowledge that is perfect and distinct can be further divided into knowledge that is perfect absolutely and knowledge that is perfect only within a genus. One is the purview of the natural philosopher, the other the purview of, for example, a master craftsman, who has perfect and distinct knowledge of copper making but does not have the perfect knowledge of all metals that a natural philosopher has. Next, there are the different meanings of "more known." The relevant one is that something is more known than another insofar as knowing it contributes to knowing that other one. Finally, there are three ways in which knowledge of one thing can contribute to knowledge of another. Either knowledge of the first directly causes knowledge of the second, as does the middle term in a syllogism; or knowledge of the first does not cause, but is necessary to, knowledge of the second, as knowledge of animal is necessary to knowledge of horse; or knowledge of the first is merely useful, as knowledge of animals is helpful for knowing plants.

The first of these three ways that knowledge contributes to other knowledge is, of course, method, for it is by method that we draw inferences. We know the facts stated in the conclusion because we know what is stated in the premises. The second and third ways are the purview of order. We treat one thing and then another, and doing so makes it easier to learn what is presented. This distinctive characteristic, Zabarella contends in chapter 11, is part of the very definition of order: order is the disposition of the parts of a discipline that make it easiest for us to learn that discipline. And, Zabarella insists, that order is always from universal to particulars, just as Aristotle said at the beginning of the *Physics*.

By appeal to the distinction between distinct and confused, Zabarella rejects the contention that universals are not better known than particulars. He says this contention is true with regard only to our confused knowledge. With regard to our distinct knowledge, universals are indeed better known. Presumably this means that for those whose knowledge is distinct, universals are indeed better known. It is only for people whose knowledge is confused that particulars are more easily and better known. Aristotle, Zabarella contends, was speaking about distinct, not confused, knowledge and so was correct to say that the order of teaching must be from universals to particulars.

One might think this is the same as to proceed from causes to effects, that is, to proceed in "compositive" order, but in chapter 12, Zabarella rejects this assumption. Aristotle, after all, he says, proceeded from effects to causes in several of his works, including the *Prior* and *Posterior Analytics*. Zabarella will return to explain this in Book II.

But before he does, Zabarella admits he must face a significant objection. Euclid in his *Elements* and Aristotle in his biological works seem to have proceeded from particulars to universals. In chapter 14, Zabarella addresses this concern. The chapter is long

and involved, but Zabarella is convinced that, contrary to first appearance, Euclid and Aristotle did in fact proceed from universals to particulars, just as he says they should have.

Book II

In Book II, Zabarella considers the three types of order proposed by Galen of Pergamum (129?–ca. 199/ca.216 CE), the ancient physician who had become as authoritative in Renaissance medical training as Aristotle had become in philosophical education. Zabarella rejects Galen's definitive order outright. He allows a place for resolutive order — the order from effects to causes — only in the practical arts. And he insists that in teaching natural philosophy, compositive order — that from causes to effects — is the only correct order. Along the way, Zabarella criticizes the newer methods being used in medical education and even medical research.[4]

Chapters 3, 4, and 5 of Book II are dedicated to condemning definitive order. Its proponents argue that definitive order completes a triad. One order begins from the beginning, one from the end, and definitive begins from the middle. Zabarella says it is simply not true that order must begin from one of these. And to say that one order alone begins with definition is wrong, too, Zabarella insists: all orders do. Or more accurately, none do, since those initial definitions are always preliminaries that lie outside of a subject's treatment, properly speaking. Zabarella proceeds to rebut several more arguments for definitive order. He concludes they are puerile, ridiculous, empty — there simply is no such thing as definitive order.

The crucial chapter in Book II is chapter 6. Here Zabarella lays down his position regarding how many possible orders there are and when each should be used. He claims his view has not been advocated by anyone since Aristotle. His position is as follows: We inquire after knowledge either for its own sake or for the sake of some other end. The first is the case in the contemplative, or

speculative, disciplines, such as natural philosophy; the second in all other disciplines, including ethics and the practical arts. When knowledge is for its own sake, we must proceed from beginning-principles (*principia*). When knowledge is for some end, we proceed from a notion of that end and continue to the beginning-principles. Zabarella finds it remarkable that philosophers and physicians have not seen this, and he commits a chapter to rebutting anyone who would think natural science can be taught by resolutive order.

Chapter 8 is an important digression. *On Methods* is about teaching, and Zabarella has argued that natural science cannot be properly taught by proceeding from effects to causes. But, he asks, might natural science be *discovered* by proceeding in this order? Could it be that a natural philosopher uses resolutive order in discovering some science and then uses compositive order when teaching it to others? Zabarella sharply rejects this and presents several arguments against it. His first is that, although there is a sense in which a natural philosopher may be said to proceed from effects to causes, this resolution is not resolutive order but resolutive method, a procedure Zabarella will explain in Books III and IV and in *On Regressus*. If the intended sense was that the natural philosopher observes lower-level species, seeks their natures and causes, and thus proceeds to high-level species, this too is not resolutive order. For the more universal is known earlier and more easily than the less universal. We do not know horse first as horse; we know it first as a natural body, then as an animal, and only finally as horse. Moreover, Zabarella continues, if the proposal is about perfect and distinct *scientific* knowledge, then it is not about any contribution from order, for all scientific knowledge is gotten by syllogistic inference, and that is the purview of method, not of order. Also, it cannot be proposed that the nature of lower-level species is scientifically known first. To know the nature of man, for example, one must have knowledge of body and soul, matter

and form. So we must know the more general before the more specific. Besides, it is just consonant with reason that Aristotle would convey his knowledge to us in the same order in which he discovered it. It would be deceptive on his part if he did otherwise. "It is, therefore, manifest that compositive order alone is appropriate both for conveying and discovering contemplative sciences." (*Meth. l. 2 c. 8 par.* 10)

Zabarella then dedicates five chapters, over a third of the whole second book, to defending and explicating his proposal with reference to the writings of ancient and proven authorities, especially to the medical works of Galen, Avicenna, and Averroës. He examines their works section by section. He compares the authors to each other. He shows where and how they used the proper compositive order and confesses he would need to censure them if their apparent deviations were truly such.

As he examines Galen, Avicenna, and Averroës, Zabarella lays down the proper relationship between the physician and the natural philosopher: Medicine is not a contemplative, natural science, but an effective art directed, as Galen, Avicenna, and Averroës all said, to conserving health when it is present and restoring it when lost. So the physician must use resolutive order, not compositive. He must accept from the natural philosopher knowledge of universal causes — knowledge of the four elements, their primary qualities, the humors, the nature of sickness and health, humoral imbalance as the cause of disease, and the natures, roles, and constituent tempers of the body's members — in short, Zabarella says, everything that falls under that first part of medicine, the part called *physiologia*. It is not fitting for the physician to debate these. His goal is practical, to maintain and restore health, so the beginning-principles that he seeks are remedies. In seeking the remedies and teaching their application, he may use resolutive order and proceed from effects to causes. He may study bodily signs.

He may use dissection. But these are tools of art and not of science. The physician must accept scientific principles from the philosopher. He cannot discover them himself.

After an extended tribute to the excellence of Galen, Avicenna, and Averroës — and to the authority of Aristotle again, and of Plato, in chapter 15 — Zabarella provides more mature definitions of both compositive and resolutive orders. He reminds his reader that neither are inferential procedures; any inference is provided by method, not by order. And he reminds the reader that, as any order must, compositive and resolutive orders both proceed from universals to particulars. No correct teaching can proceed otherwise.

Book III

Book III is the only one dedicated specifically to method. Zabarella claims that there are only two legitimate types — demonstrative and resolutive. The first infers cause from effect; the second infers effect from cause.

In Books I and II, Zabarella insisted that it is method, not order, that moves us from knowledge of what is known to knowledge of what is not yet known. In Book III, he repeats this and adds that the only logical structure that can effect this movement is the syllogism. Even induction, enthymeme, and example are effectual only when they can be reduced to the syllogism. Syllogism and method, therefore, are the same, at least in form. The only difference, Zabarella explains in chapter 3, is that in method, the premises are known to be true, whereas premises in dialectical and oratorical syllogisms need not be.

Chapter 4 contains the gist of Zabarella's theory of method. He says there are two types. The first and stronger type argues from cause to effect and is commonly called demonstration *potissima* or demonstration *propter quid*. Zabarella calls this demonstra-

tive method. The lesser of the two types argues from effect to cause and is commonly called demonstration *a signo* or demonstration *ab effectu*. Zabarella calls this resolutive method. The two types of order were compositive and resolutive, but the two types of method are demonstrative and resolutive. A reader must take care to note the difference and also that (as Zabarella usually speaks) both demonstrative method and resolutive method are types of demonstration. There is, Zabarella insists, no "scientific method" besides these two, "no other instrument for knowing scientifically" (*Meth. l.* 3 *c.* 4 *par.* 6). Their natures and roles are revealed incidentally in the polemical chapters that follow and then described explicitly in *On Regressus*.

The first object of Zabarella's attack are those who treat division as a method.[5] By this method, they say, we come to know, for example, man by first dividing all bodies into animate and inanimate, then all animate into sentient and not, then all sentient into rational and not, and thus come to know man as the rational animal. This may be a useful order in which to teach, Zabarella says, but it is not method. Consider using this procedure to find the nature of a zoophyte, he suggests. When faced with the question of whether it is sentient or not, a process of division will be of no help. One is similarly at a loss when the number of species in a genus is unknown. For these, one needs method, a way of proceeding from what is known to what is unknown; that is, one needs syllogistic inference, and division is not such. Division, that is, is not method.

A procedure that recursively infers one thing from another must, of course, begin somewhere. For Zabarella, it begins not with those things that are known by method, but with things known all by themselves, things known *per se*. These are not the primary subject of *On Methods*, but we get a glimpse of what Zabarella means in chapter 8. He says that what is known *per se* is

of two types. The first is that which is most evident and actually known by everyone. Axioms are an example. Second are things that may not be known by everyone but are readily accepted once they are heard and considered a little. The definition of a circle is an example.

Zabarella begins his attack on definitive method in chapter 11. Like his dismissal of divisive method, Zabarella's dismissal of definitive method is lengthy but essentially simple: definition is not a method because it is not syllogistic inference. Across five chapters, Zabarella recounts and dismisses many alternative views. He must have had many opponents on this, but he does not identify them. As he proceeds, we again hear that what is learned without method, such as whether a particular attribute is essential to a substance, is learned by turning the question over in one's mind — like turning a log in the fire and seeing what was earlier unapparent (*Meth. l.* 3 *c.* 14 *par.* 6).

In chapter 16, Zabarella corrects what he considers misunderstandings about resolutive method. He rejects, for example, the proposal that it is a proceeding from particulars to universals. Any such proceeding would not be method, he insists, since method is a proceeding from known to unknown and the universal is always known first and better. Even coming to know, for example, the man Socrates is coming to know him first as a bodily thing, then as man and only lastly as a particular man.

Zabarella closes the book with restatements and summaries of what he has said and then discussions of two miscellaneous topics. The first is whether the methods can be further subdivided. He says that demonstrative cannot but resolutive has two types. One is demonstration *ab effectu* (or *a signo*). The second is induction. The first, he says, is the more important and more effectual and the one used for discovery of what is highly obscure and hidden. It is so important in fact that throughout *On Methods*, Zabarella of-

ten treats it as synonymous with resolutive method. Induction, he explains, is much weaker and is used for discovery of those things that are not completely unknown and need only a light clarification or confirmation. By it we may know that all men are biped. But for discovery of the abstract beginning-principles of science, it is "utterly useless" (*Meth. l.* 3 *c.* 19 *par.* 6). Finally, Zabarella notes similarities between resolutive method and resolutive order but reminds the reader that resolutive order is used only for the teaching of practical disciplines, where, strictly speaking, no method of any kind is used.

Book IV

Zabarella has argued that method, that is, syllogistic inference from known premises, is the only instrument that can lead to scientific knowledge, and for this theory he has taken Aristotle's logical works as his primary authority. This creates a problem. The second book of the *Posterior Analytics,* the culmination of what Zabarella considers Aristotle's logical works and a book of central importance in sixteenth-century natural philosophy, appears to be about using definition to acquire scientific knowledge. In Book IV, Zabarella addresses the apparent contradiction.

Zabarella says there have been two schools of thought regarding what Aristotle is doing in the second book of the *Posterior Analytics.* The older view is that the book is about the discovery of the middle term in a demonstration. If the book is about definition, it is about definition as that middle term. The newer view is that the book is indeed about definition as an instrument for securing scientific knowledge, an instrument distinct from method. The first five chapters of *On Methods,* Book IV, recount and rebut the first position, the next three chapters the second position. The remainder of the book presents Zabarella's own interpretation, an interpretation he claims is new.

Zabarella insists that all of the *Posterior Analytics,* including the second book, is directly or indirectly about demonstration *potissima.* Many commentators—before, in, and after Zabarella's time —think Aristotle's intent in the work is to develop an inductive or empirical theory of science, a theory whose presentation culminates in the discussion of induction in the second book's final chapter, our chapter 19. Zabarella disagrees, and he believes the bulk of that chapter is an appendix incidental to Aristotle's main project. To the extent the second book is about definition, he says, it is about how demonstration provides us with definitions, especially how demonstration *potissima* provides us with definitions of accidents.

While criticizing opposing views, Zabarella recalls Aristotle's claim that a definition is (in Zabarella's words) the beginning principle of a demonstration, the conclusion of a demonstration, or the very demonstration itself, differing in position (*Meth. l.* 4 *c.* 4 *par.* 7). Zabarella says recent interpreters have concluded from this that substances are known by definition and accidents by demonstration. But both, he says, are known by demonstration. Or specifically: accidents must be and are known by demonstration, substances need not be but often are also.

It is because they have causes external to themselves that accidents must be and are known by demonstration, specifically by demonstration *potissima,* that is, demonstrative method. But substances do not have external causes on which they depend and so need no demonstration to be known. They can be known "by means of themselves" (*per se*) or "from themselves" (*ex se*). We, however, because of the feebleness and inadequacy of our wits, often use demonstration to know them. When we do, we do not use demonstration *potissima,* demonstrative method, but instead use resolutive method. Zabarella explains how we do so in chapter 13, and then in chapter 14 how we use demonstrative method to define accidents.

There are two steps, Zabarella explains, in coming to know the definition of a substance. The first uses division and the second uses induction. To obtain the definition of some genus, we identify all the essential accidents of each of the genus's species and exclude the accidents that are essential to each particular species; the essential accidents that remain are essential not to any particular species but to all of them. All that is the first step and it is provided, Zabarella says, by a process of division. The result is the list of accidents that are predicated as essential to all the species. The second step uses induction, which for Zabarella is a kind of demonstration: what is essential to all the species is essential to the genus. Because induction is one of the two types of resolutive method, Zabarella can say that the definition of a substance is the "unknown end of resolutive method," but he stresses that both steps are required. He closes by noting that in some cases, the other type of resolutive method, demonstration *ab effectu*, or *a signo*, can substitute for division. In either case, it is the first step, not the induction, that identifies what is universal among the particulars. For Zabarella, induction itself is not ampliative.

But if the second book of the *Posterior Analytics* is about definition at all, it is primarily about definition of accidents not substances, Zabarella says repeatedly, and he treats such definitions in chapter 14 of *On Methods*, Book IV. He begins by again recalling Aristotle's three types of definition—the beginning principle of a demonstration, the conclusion of a demonstration, and a demonstration differing by position. Zabarella criticizes his predecessors for misunderstanding the distinction, and to understand Zabarella we must be careful not to confuse his interpretation with theirs.

The perfect definition of an accident, Zabarella says, identifies matter, form, and cause. For accidents, the genus is the form, and the differentia is the matter. In the definition of an eclipse as privation of the light of the moon, privation of light is the form, moon is the matter. Zabarella calls this definition essential, quidditative,

and nominal, but also deficient and imperfect, because it fails to identify a cause. A complete, or perfect, definition would be privation of light of the moon caused by interposition of the earth. A perfect definition can be cast as a demonstration. The genus is the major term, the matter is the minor term, and the cause is the middle term: interposition of the earth causes privation of light; the moon suffers an interposition of the earth; therefore the moon suffers a privation of light.

Note that the conclusion of the demonstration does *not* include all the terms of a perfect definition. In chapter 15, Zabarella criticizes his predecessors for thinking so. The conclusion provides merely a nominal definition. It includes genus and species but not cause. The perfect definition comprises all three terms of the demonstration, simply rearranged; such a definition is a demonstration differing in position. The third type is the cause alone, without reference to genus or differentia, to form or matter. It is this cause that Zabarella calls the beginning-principle of the demonstration. This beginning-principle is *not* a premise or any other proposition. It is a term in a proposition. Here it is the interposition of the earth. Zabarella criticizes his predecessors for their confusion about this too.

In chapter 16, Zabarella completes his presentation of the relationship between demonstration and definition by identifying the priority. It is by coming to know the demonstration *potissima* that we come to know the definition. It is not the other way around. For accidents, the end (not the conclusion) of demonstrative method, that is of demonstration *potissima*, is definition. We come to know the demonstration, then we rearrange the terms to get the definition. The definition "comes to light by means of" the demonstration. We can come to know the definition of an accident in no other way.

The final chapters defend Zabarella's theory by reference to Aristotle and Averroës and further criticize Zabarella's opponents.

ON REGRESSUS

In the short work *On Regressus*, Zabarella explains regressus, defends it against critics, gives examples, and explicates its most elusive step.

On Methods was about teaching. It left unanswered the important question of how we come to have distinct scientific knowledge, that is, how we come to know a demonstration *potissima*. In *On Regressus* Zabarella assumes the reader knows of the proposal that we do this by a process called regressus. He also acknowledges that the process has been criticized as a form of circular reasoning, and he responds to the criticism in the work's first two chapters. He admits a superficial similarity. Both demonstrate B from A and then turn around and demonstrate A from B. But, Zabarella claims, the whole argument in a regressus is not circular, because the two demonstrations are qualitatively different. The first is a demonstration *quòd*, and the latter is a demonstration *propter quid*. Zabarella explains this difference in chapters 3 and 4.

The first procedure infers cause from effect. We see smoke and infer there is fire. We see that planets do not twinkle as the stars do, and we infer that the planets must be close. Such a demonstration *ab effectu* (from effect) is a demonstration *quòd*, a demonstration that something is the case. We know smoke implies fire, but we do not know fire *as* the cause. Our knowledge, Zabarella says, is confused.

How, critics ask, do we even know that smoke implies fire? Zabarella answers: by the process Averroës called demonstrative induction. He uses this example: We see that changes in accidents have an underlying subject, but we do not see this with substances. But after we consider this a little, we recognize that an underlying subject must be essential to all changes, not just changes of accidents. By this insight, we conclude that what is true of the observed particulars is, by essence, true of all particulars. We then

perform the final step, the inductive step: what is true of all is true of the class. We thus get the elements of the demonstration *quòd*: where there is generation (a kind of change), there is underlying subject matter; in natural body there is generation; therefore, in natural body there is underlying subject matter. To get distinct knowledge, however, we need to turn this demonstration around, so as to infer effect from cause, not cause from effect.

But an intermediate step is first required. Zabarella calls it a mental consideration. He claims that no one before him has explained how it is performed and gives his own explanation in chapter 5. First, he says, we recognize the constant conjunction. Then, by comparing the two things conjoined, "it happens that we are led little by little to knowledge of characteristics of the [cause] . . . until finally we know that this is the cause of that effect." Zabarella makes no mention of experimental research or additional observation. For him, this part of the process is purely one of progressive cognitive insight. By it we come to know the cause distinctly.

When the cause is known distinctly, its nature is known; it is recognized *as* the cause, and the final demonstration can be formed as follows: where there is underlying subject matter, there is generation; in natural body there is underlying subject matter; therefore, in natural body there is generation. We now have a demonstration *propter quid*, an inference of the effect from the cause. We can infer smoke from fire, generation from underlying matter, lack of twinkling from nearness of planets. We now have what in *On Methods* was called a demonstration *potissima*. By virtue of the mental consideration that formed the middle step in a regressus, we have distinct scientific knowledge.

In chapter 6, Zabarella gives another example, this one from the eighth book of Aristotle's *Physics*, by which we come to distinct knowledge of the first mover as the first cause of motion. In chapter 7, Zabarella notes that the three steps of a regressus are always

logically separate, even when they occur at the same time. In the final two chapters, 8 and 9, Zabarella returns to distinguishing regressus from reasoning in a circle and to further rebutting his opponents.

ZABARELLA'S OPPONENTS AND SPIRIT

On Methods and *On Regressus* are polemical works against unnamed opponents. The first two books of *On Methods* are directed against teachers and writers of scientific subjects who order their presentation by treating particulars first and then universals, instead of vice versa. The latter two books are primarily directed against those who promote a kind of order they call definitive; Zabarella insists there is no such thing. *On Regressus* is directed against those who dismiss regressus as circular reasoning.

Zabarella does not name his opponents, but he identifies them as errant natural philosophers, especially — and particularly in the first two books — those who nowadays (*in his temporibus*) teach medicine. They have abandoned the order used by proven authors (*probati authores*) of the past, including Aristotle, Galen, Avicenna, and Averroës. Zabarella continually defends his arguments by citing this group of established authorities. In *On Methods* and *On Regressus*, we do not see a writer at the vanguard of a scientific revolution. We see a conservative standing athwart history and yelling, "Stop."[6] He does not present in these books a prototype of Baconian induction or Galilean experimentalism.[7] If Zabarella anticipates any major figure in the emergence of modern science, it is Descartes. For both, scientific knowledge can be obtained only by deductive inference from premises made clear and distinct using only cognitive reflection and insight.

At times, Zabarella is not only conservative but positively retrograde. An example is his understanding of induction. Since the Alexandrian synthesis of Aristotelianism and Neoplatonism

crafted in late antiquity, induction was understood as a defective form of syllogistic inference that could be made good only by an enumeration that was (actually or supposedly) complete. Zabarella adopts and uses this conception in *On Methods* and *On Regressus*, but also in his textbook *Tabula Logicae* and his commentary on the *Posterior Analytics*. In the summary above, we see Zabarella use induction to infer that something is true of a class if it is true of all members. For him, any inference from some to all happens outside of induction. And Zabarella thought this was the correct view of induction because, he believed, it was Aristotle's view. But this stock interpretation had come under attack as early as Jean Buridan (ca. 1300–1358/61) and was increasingly challenged as humanist scholars discovered a different conception of induction in Cicero and in Aristotle's *Topics*. By 1542, a commentary by the Paduan Agostino Nifo could say that there was vexing disagreement over what induction was.[8] By the time Zabarella began publishing, it had become commonplace that, at least in natural philosophy, there was a kind of induction that could be used to obtain definitions and thus scientific knowledge. Zabarella gives this more up-to-date conception of induction no attention — unless it is the object of his attack on those championing definitive method.

Zabarella was the last significant professor of natural philosophy to work within the conceptual and cultural framework of Aristotelian commentary, the last whose magnum opus could be an extended commentary on the *Posterior Analytics*. Zabarella died in 1589, in the same city in which he was born, was schooled, and had taught for thirty-five years. Interest in his work diminished rapidly. The last edition of *On Methods* and *On Regressus* was that of 1623, the year Galileo published *The Assayer* and a year in which Descartes was shifting his attention from mathematics to scientific method. Whatever ideas Zabarella may have bequeathed to his successors would need to appear inside new conceptual frameworks.

NOTES

1. On Zabarella's life and works, see Mikkeli, "Giacomo Zabarella." This article has an extensive bibliography.

2. Randall, "The Development of Scientific Method."

3. Francis Bacon, *The Proficience and Advancement of Learning* (London: Henrie Tomes, 1605), *l. 2 c.* 16–17, spelling modernized.

4. Lawrence I. Conrad, Michael Neve, Vivian Nutton, Roy Porter, and Andrew Wear, *The Western Medical Tradition, 800 BC to AD 1800* (Cambridge: Cambridge University Press, 1995), 253–55. For the context in which Zabarella was writing and his likely opponents, see also Andrew Cunningham, "Fabricius and the Aristotle Project in Anatomical Teaching and Research at Padua," in *The Medical Renaissance*, ed. Wear, French, and Lonie.

5. Regardless of how much Zabarella himself had Peter Ramus in mind on this, the Paduan's writings did get used to attack Ramus and his many followers and imitators. See Maclean, "Mediations of Zabarella," and especially the dedication cited therein to Giulio Pace's 1586/87 edition of Zabarella's logical works.

6. William F. Buckley's description of a political conservative.

7. For an opposing assessment, see Wallace, "Circularity and the Paduan Regressus," who argues for the indirect influence of Zabarella on an early logical work of Galileo.

8. Agostino Nifo, *Aristotelis Stagiritae Topicorum libri octo* (Venice: Girolamo Scotto, 1557), f. 18v.

DE METHODIS

ON METHODS

LIBER PRIMUS

Praefationem continens.

1 Omnem scientiam, omnem artem, omnemque disciplinam methodo aliqua tradi et absque methodo consistere non posse manifestum est. Nullam enim facultatem benè docere aliquis potest, nisi et in illius partibus disponendis et in singulis theorematibus declarandis, tradendaque rerum absconditarum cognitione methodum aliquam servet.

2 Hinc ortum habuit logica disciplina, quae tota in methodorum traditione constituta est, proinde instrumentalis facultas iure nuncupatur, quoniam omnibus disciplinis instrumenta, id est methodos praebet, quibus ad rerum notitiam° adipiscendam iuvemur.

3 Propterea nulla ad logicum magis pertinens, nulla utilior tractatio esse potest, quàm de methodis, de quibus dicere in praesentia constituimus.

4 Scripserunt quidem hac de re complures tum antiquiores tum posteriores philosophi, necnon et medici, dum[1] occasionem huius tractationis à Galeno sumpserunt in principio Artis medicinalis.

5 Qui omnes licèt in quibusdam parvi momenti discrepare videantur, in eo tamen, quod praecipuum est, ita unanimes et concordes extitere, ut his temporibus nemo aliam sententiam proferre aut excogitare ausus sit. Cùm enim methodum latè sumptam in methodum propriè acceptam et ordinem partiantur, quatuor esse methodos dicunt, demonstrativam, resolutivam, definitivam ac divisivam; ordines autem tres, compositivum, resolutivum, definitivum, authore Galeno in eo loco.

BOOK ONE

: I :

Containing the preface.

It is manifest that every science, every art, and every discipline is 1
conveyed by some method and cannot endure without method.
For in no branch of learning can anyone teach well, unless both in
disposing its parts and in clarifying every theorem, and in conveying knowledge of hidden things, he maintains some method.

And thus, the discipline of logic, which is wholly constituted in 2
the conveying of methods, came to be. Accordingly, it is justly
called the instrumental faculty, since it provides instruments for all
disciplines, that is, methods, by which one may be helped in obtaining knowledge° of things.

And so there can be no treatment pertaining more to logic, 3
none more useful, than one regarding methods, about which we
have decided to speak at present.

Of course, many philosophers, both ancient and later, wrote 4
about this issue. And so too did physicians — when they took the
opportunity for [writing] such a treatment from Galen, at the beginning of *The Art of Medicine*.[1]

Granted that they all appear to differ in some things of little 5
moment, nevertheless in that which is principal they were so
unanimous and like-minded that these days no one has dared to
advance or imagine another position. Since they divide method
taken in a broad sense into method accepted properly and order,
they say there are four methods — demonstrative, resolutive, definitive, and divisive — and three orders — compositive, resolutive,
and definitive — following the author Galen in that passage.

3

6   Quae quidem communis omnium sententia si ita vera sit, ut omni difficultate careat, nos de ea aliquid scribere operae pretium non esset, quandoquidem tam multa ab aliis scripta ad huius dogmatis declarationem comperiuntur, ut praeterea nihil nobis de eadem re dicendum relinqueretur. Alienum autem à nostro instituto penitus est illa commemorare, quae apud alios multos scriptores legi possint.

7   Verùm ego de hac communi opinione semper dubitavi et eò magis dubitavi, quò eam diligentius consideravi atque animo volui. Tandem verò cùm in iuventute logicam publicè profiteri et logicos Aristotelis libros assiduè tractare coepi, in aliam sententiam devenire coactus sum, in qua usque ad hodiernam diem perseveravi. Falsum enim esse puto aliorum dogma de methodorum atque ordinum numero. In hoc autem si decepti sunt, in cognoscenda quoque tam methodi quàm ordinis natura eos aberrasse necesse est. Cuius erroris (ni fallor) causa[2] fuit, quòd professores logicae, qui in Aristotelis libris interpretandis versantur, rem hanc, quae ad eos maximè pertinebat, non consideratunt, sed penitus praetermiserunt. Medici verò, quorum id non intererat, soli hoc munus sibi arrogarunt; quo factum est ut morborum et pharmacorum consideratione impliciti, et librorum Aristotelis non satis memores, logicam quaestionem enodare nequiverint.

8   Hac igitur de re statui aliqua scribere, haud equidem altercandi studio, neque quòd putem me unum in hac difficultate scopum attigisse, quem Galenus et alii multi viri clarissimi assequi non potuerint; sed solo veritatis amore ductus. Ita enim fiet ut vel alii in meam sententiam veniant, si veram[3] esse cognoverint°; vel, si ego deceptus fuero, saltem ea dubia proferens, quae hactenus me à

Of course, if this position, common to all, were indeed so true 6
that it was free from all problems, it would not be worth the work
for us to write anything about it, since so much written by others
for clarification of this doctrine can be found that nothing in addition would remain for us to say about it. And moreover, it is completely alien to our intent to rehearse that which may be read in
many other writers.

But in truth, I always had doubts about this common opinion, 7
and I doubted it all the more, the more I carefully considered it
and turned it over in my soul. And when finally, in my youth, I
began to lecture on logic publicly and to treat Aristotle's logic
books assiduously, I was forced to come to another position, and I
have persevered in it down to this very day. For I now hold that
others' doctrine about the number of methods and orders is false.
And furthermore, if they were deceived in this, it is necessary that
they also erred in [their] knowledge of the nature of method as
much as [the nature] of order. The cause of the error, unless I am
mistaken, was that professors of logic, those concerned with commenting on Aristotle's books, did not consider this issue; it per-
tained especially to them, but they completely overlooked it. And
physicians, to whom it was of no concern, arrogated this job to
themselves alone; hence it happened that they, involved as they
were in consideration of diseases and medicines and not mindful
enough of Aristotle's books, were unable to untangle a logical
question.

I decided, therefore, to write some things about this, not of 8
course out of a keenness for wrangling, or because I hold that,
with this problem, I alone have reached a goal that Galen and
many other illustrious men were unable to secure; rather, I was
led only by the love of truth. Now it will happen that either others will come around to my position, if they recognize° that it
is true, or, if I have been deceived, then at least by advancing
doubts that have up until now kept me away from the positions of

Galeni et aliorum sententia retraxerunt, aliquem excitem veritatis amatorem, qui animum meum dubiis solvat, meque in errore diutius esse non patiatur. Profiteor enim me non modò veritatem toto animo amplexaturum, verùm etiam ei, qui in me tale beneficium contulerit, plurimum debiturum. Quemadmodum etiam de omnibus bonarum artium studiosis, iis praesertim, qui ingenuo animo praediti sunt, illud mihi ipse polliceor, si per me veritatem cognoverint,° aut saltem ad eam indagandam aliquid è meis scriptis auxilii luminisque susceperint, eos mihi laboribusque meis mediocres gratias esse habituros.

: II :

Quid sit methodus latè accepta.

1 De methodis atque ordinibus locuturi ante omnia quid sit commune horum genus Methodus intelligere debemus, nam, si vocis significationem spectemus, non modò illas, quas propriè methodos vocant, sed ordines quoque, quos à methodis separant, nominare methodos possumus, etenim viae quaedam sunt et transitus ab aliquo ad aliquod, quae est Graecae vocis, μέθοδος, propria significatio.⁴

2 Haec igitur amplè sumpta methodus nil aliud esse videtur, quàm habitus logicus, sive habitus intellectualis instrumentalis nobis inserviens ad rerum cognitionem adipiscendam. Non enim omnis via et omnis transitus solet methodus appellari, sed quae à mente nostra fit scientiam rerum investigante.

3 Quoniam autem Methodus est habitus logicus, non est ignorandum, duplicem esse Methodum, quemadmodum et duplicem esse logicam aliàs declaravimus in eo libro, quem de natura logicae

Galen and others, I might excite some lover of truth, who can re-
lease my soul from [its] doubts and let me be in error no longer.
For I profess, not only would I then embrace the truth with all my
soul, but I would also be greatly obliged to him who conferred on
me such a benefit. And in the same way, regarding all those studying the humanities, especially those who are endowed with a noble
soul, I assure myself [of] this: if through me they should recog-
nize° the truth, or at least take from my writings something of aid
and light for tracking it down, they will have a modicum of grati-
tude for me and my efforts.

: II :

What method, accepted in a broad sense, is.

In speaking about methods and orders, we ought, before all else, 1
to understand what Method,[2] their common genus, is. For if we
look at the signification of the word, we can name as methods not
only those things that they properly call methods, but also orders,
which they separate from methods. For indeed these are as it were
ways and passages from something to something, which is the
proper signification of the Greek word, *methodos*.

This method therefore, taken in the wide sense, appears to be 2
nothing other than a logical habit,[3] or an instrumental habit
of [the] understanding that serves us in obtaining knowledge of
things. For normally not every way and every passage is called a
method, but only that which is taken by our mind as it investi-
gates the science of things.

Moreover, since Method is a logical habit, it should not be ig- 3
nored that Method is of two types, just as also, elsewhere, in the
book that we wrote *On the Nature of Logic*,[4] we made clear that

scripsimus, ex quo multa sumi possunt ad praesentem tractationem conferentia. Ibi quidem logicam duplicem esse diximus, unam rebus applicatam et iam in usu positam, alteram à rebus seiunctam. Hanc scientiam non esse, sed scientiarum instrumentum seu instrumentalem disciplinam demonstravimus; illam verò non amplius instrumentum, sed scientiam, fit enim scientia illa, cui dicitur applicata. Eadem ratione methodum appellare possumus tum ipsum instrumentalem habitum nulli adhuc rei seu disciplinae applicatum; tum etiam disciplinam ipsam, cui est applicatus. Propterea Aristoteles multis in locis scientias ipsas ac disciplinas methodos appellare solitus° est, nam methodo utuntur omnes, methodoque consistunt; siquidem nihil aliud sunt, quàm logicae methodi in usu positae. Nos autem in praesentia methodum non rebus applicatam sed à rebus abiunctam considerandam proposuimus. De methodo autem mathematica vel naturali verba facere esset de rebus ipsis ac de ipsarum scientiarum natura disserere, quod sanè à nostro instituto esset prorsus alienum, qui solam ipsius methodi naturam volumus contemplari, ideò eam diximus esse habitum intellectus instrumentalem, quo ad rerum cognitionem consequendam iuvamur.

4   Cùm igitur methodus sit instrumentum, qua differentia ab omnibus corporeis instrumentis discrepet manifestum est, non est enim corporale instrumentum, sed intellectuale et habitus quidam animi nobis inserviens pro aliorum principalium habituum adeptione.

5   Hinc colligitur etiam differentia maximè quidem propter illa, quae dicenda sunt, à nobis notanda inter methodum et aliud quiddam instrumenti genus, cuius mentionem fecit Aristoteles in principio libri de Interpretatione in capite de oratione, ubi haec verba leguntur, 'Est autem omnis oratio significativa non sicut instrumentum, sed ex placito et arbitrio humano.' His verbis carpit Aristoteles Platonem, qui putavit omnia nomina et verba et ex his

logic is of two types. From that book many things that contribute to the present treatment can be gotten. We said there, of course, that logic is of two types — one is applied to things and thus put into use, the other is separate from things. We demonstrated that the latter is not a science but an instrument of the sciences or an instrumental discipline, and the former is no longer an instrument, but a science — for it becomes a science in that to which it is said to be applied. For the same reason we can call method both that instrumental habit not yet applied to anything or to any discipline, and also the very discipline to which it is applied. Aristotle therefore in many passages is wont° to call the sciences and disciplines themselves methods, for they all use a method and consist in a method, since they are nothing other than logical methods put into use. In the present [work], however, we have proposed to consider method not applied to things but separated from things. To offer up some words on mathematical or natural method would be to discuss the nature of things and the nature of the sciences themselves, and that would surely be utterly alien to our intent. We want to contemplate only the nature of method itself. Because of this we said it is an instrumental habit of [the] understanding by which we are helped in gaining knowledge of things.

While method therefore is an instrument, by what differentia it 4 is different from all corporeal instruments is manifest; for it is not a corporeal instrument but one of [the] understanding and some sort of habit in man's soul, serving us in obtaining other important habits.

Now from this is gathered a difference between method and 5 any other kind of instrument. It has to be noted by us especially, of course, because of what will be said. Aristotle made mention of it in the beginning of the book *On Interpretation* in the chapter on speech, where these words can be read: "All speech is significant not as an instrument but from human preference and choice."[5] With these words Aristotle criticized Plato, who held that all

conflatam orationem esse instrumenta naturalia, id est à natura nobis tradita ad res significandas, prout cuiusque rei propria natura propriam et sibi convenientem appellationem postulabat.

6   Aristoteles autem hoc negat, quia cuiusque rei propriam nominationem ex humano tantùm arbitrio, non ex natura rerum pendere existimavit. Possumus tamen fortasse hos philosophos conciliare, si dicamus vocem significatricem duas habere partes, unam veluti materiam, alteram veluti formam. Materia quidem est sonus, qui à natura est; forma verò articulatio, quae tota à nostro arbitrio pendet. Sed an hoc Plato voluerit non est in praesentia considerandum. Utcumque enim hoc sese habeat, illud certum esse debet, Aristotelem non negare voces articulatas instrumenta esse significandi conceptus animi, sed negare esse instrumenta naturalia eo modo, quo arbitrabatur Plato. Nam revera instrumenta quaedam sunt, quibus ea, quae animo concipimus, significamus, non tamen sunt methodi, neque instrumenta logica. Nam methodus est habitus animi, non est vox articulata; et est instrumentum non ad significandum sed ad scientiam comparandam inventum.° Praeterea omnis methodus et omne instrumentum logicum est via et processus ab aliquo ad aliquod, quae sint eiusdem generis seu eiusdem ordinis, ut à re ad rem vel à conceptu ad conceptum vel à voce ad vocem, haec enim omnia verè dici possunt, logicus namque discursus potest et in rebus et in conceptibus et in vocibus considerari, isque momento temporis fieri non potest, sed cum aliqua mora et in tempore.

7   At dum per vocem significatur res vel eius conceptus, non fit progressus inter illa, quae sint[5] eiusdem ordinis, sed à voce ad rem seu conceptum significandum, idque fit subitò et in temporis momento. Ideò neque processus neque discursus propriè dici potest,

nouns and verbs and speech assembled from them are natural instruments, that is, conveyed to us by nature for signifying things,
in that the proper nature of each thing demands an appellation
that is proper and appropriate to that thing.[6]

Aristotle denies this, because he judged that the proper naming   6
of anything depends only on human choice, not on the nature of
things. Nevertheless, perhaps we can reconcile these philosophers,
if we say that a word, as signifier, has two parts — one as it were
material, the other as it were formal. The material, of course, is
the sound, which is by nature, while the form is the articulation,
which depends wholly on our choice. Now whether Plato would
have accepted this does not at present have to be considered. For
however that may be, it ought to be certain that Aristotle did not
deny that articulated words are instruments for signifying concepts in man's soul but denied that they are natural instruments
the way Plato thought they were. For in truth they are a sort of
instrument, by which we signify those things which we conceive in
our soul. Nevertheless, they are not methods or logical instruments. For a method is a habit in man's soul; it is not an articulated word; and it is an instrument invented° not for signifying but
for procuring scientific knowledge. In addition, every method and
every logical instrument is a way and a proceeding from one thing
to another, both of which belong to the same genus or the same
order, as from thing to thing or from concept to concept or from
word to word. For all these truly can be said, and logical discursive
movement can be considered in things, in concepts, and in words;
and [as a movement] it cannot happen in a moment of time but
only with some lapse [of time] and in time.

Now when a thing or its concept is signified by means of a   7
word, progression does not occur between those things that are of
the same order but from the word to the thing or concept being
signified, and this occurs suddenly and in a moment of time. Because of this, it cannot properly be called either a procedure or a

praesertim cùm non sit ab alio ad aliud, sed potius ab eodem ad idem. Licèt enim vox materialiter sumpta differat à conceptu, formaliter tamen non differt, quia conceptum ipsum significat, quare nullus potest esse progressus, à voce significante ad rem vel conceptum significatum.

8   Quòd autem tale instrumentum significandi conceptus non sit instrumentum logicum, neque in talium instrumentorum fabricatione° versetur logicus, res per se clara et perspicua est, duas enim partes vocem habere diximus, sonum ut materiam et articulationem ut formam. De sono logicus agere non debet, quia id est munus philosophi naturalis; neque de articulatione ipsius, quia considerare an hoc vel illo modo sonus sit articulandus pro significatione rerum ad Grammaticum pertinet vel ad nominum impositores. Ideò Aristoteles in ipso statim logicae artis initio accepit tanquàm notum quòd voces sint significatrices conceptuum, neque consideravit an hoc vel illo modo formandae sint, sed ab aliquo alio artifice iam formatas sumpsit, nec ullam amplius mentionem talis instrumenti in libris logicis fecit.

9   Praeterea sciendum° est omnem methodum esse habitum logicum sive habitum intellectualem instrumentalem et omnem talem habitum esse methodum, nec propterea perfectam esse definitionem methodi, si dicamus methodum esse habitum intellectus instrumentalem, quamvis haec definitio ex genere et differentiis constet et cum definito, ut dictum est, reciprocetur. Nam docuit Aristoteles in secundo libro Posteriorum Analyticorum accidentis definitionem non esse perfectam, etiam si essentialis sit et aequalis definito, nisi causam externam exprimat, à qua accidens pendet. Est quidem essentialis haec definitio, eclipsis est privatio luminis

discursive movement, especially since it is not from one to another but rather from one to the selfsame. Granted that a word taken materially differs from a concept, it still does not differ [from it] formally, because it signifies the concept itself. And so, there can be no progression from the signifying word to the thing or concept signified.

Moreover, that such an instrument for signifying a concept is not a logical instrument and a logician is not concerned with construction° of such instruments is an issue clear and plain *per se*. For we said that a word has two parts: the sound as matter and the articulation as form. The logician ought not deal with the sound, because that is the job of the natural philosopher, or with its articulation, because to consider whether, for signification of things, the sound should be articulated in this way or that way pertains to the grammarian or to the imposers of names. Because of this, Aristotle accepted straightaway, as already known at the very start of the art of logic, that words are signifiers of concepts. He did not consider whether they should be formed in this way or that. He took them as already formed by some other practitioner and did not make any further mention of such an instrument in the books of logic.

Continuing then, it has to be understood° that every method is a logical habit or an instrumental habit of [the] understanding and that every such habit is method. But it is not thereby a perfect definition of method if we say that method is an instrumental habit of [the] understanding, even though this definition is composed out of genus and differentiae and can be reciprocated, as is said, with what is defined.[7] For in the second book of the *Posterior Analytics*,[8] Aristotle taught that the definition of an accident is not perfect, even if it is essential and equal in total extent to what is defined, unless it expresses the external cause upon which the accident depends. This definition, of course, is essential—an eclipse is the privation of the moon's light—and it reciprocates with

lunae, et cum eclipsi reciprocatur, sed non est perfecta, nisi adda-
tur causa externa,[6] quae est obiectio terrae. Quoniam igitur methodus instrumentum est, et cuiusque instrumenti causa praecipua
est ipse finis, idcirco in definitione methodi finalem causam adieci-
mus, quae est rerum cognitio à nobis per methodum acquirenda,
in hac enim consistit tota methodi natura. Quam respiciendo
possumus non incongruè methodos omnes appellare doctrinas,
hoc est modos docendi et tradendi omnes scientias et artes, ut
doctrinae nomen finem ipsum methodi significet.

10   Sed aliqui fuerunt, qui doctrinam pro genere methodorum
atque ordinum, imò et scientiarum et artium omnium acceperunt.
Hos enim omnes habitus considerarunt prout docentur, id est
prout à doctore eos habente discipulis[7] traduntur, ut si quis doceat
alios scientiam naturalem ex eius habitu, quem ipse habet, scientia
naturalis vocabitur doctrina. Quare si praeceptor habens habitum
logicae doceat alios viam ipsam procedendi, nempè ordines ac methodos, ita ordines ac methodi appellabuntur doctrinae, quia traduntur discipulis à praeceptore. Propterea in definitione ordinis et
methodi doctrinam pro genere acceperunt.

11   Quam sententiam ego nulla ratione° probare possum, quandoquidem huiusmodi consideratio particularis et accidentaria penitus
est. Et perinde faciunt, ac si de hominis natura tractationem insti-
tuentes hominem ut ambulantem vel loquentem vel alba veste in-
dutum considerarent atque eiusmodi accidentia in definitione hominis acciperent. Methodi quidem et ordinis natura ea est, ut sint
habitus animi ad scientiarum et artium traditionem nobis inser-
vientes seu omnino ad cognitionem capessendam. Quòd autem
aliquis habitu logico praeditus alios logicam doceat et aliis eosdem
habitus tradat, accidentale est et ab ipsa logicae natura alienum.
Quemadmodum arti aedificatoriae accidit ut ille, qui eius habitum
habet, eam doceat alios. Hic enim non est scopus illius artis, sed

eclipse. But it is not perfect unless the external cause, which is the interposition of the earth, is added. Since method therefore is an instrument, and the principal cause of an instrument is the end itself, we added into the definition of method a final cause, which is the knowledge to be acquired by us by means of the method. For the whole nature of method consists in this. In this regard, we can, not inaptly, call all methods teachings, that is, ways of teaching and conveying all the sciences and arts, since the name of teaching signifies the very end of method.

But there were some who accepted teaching not only as the ge- 10 nus of methods and orders but also of all the sciences and arts. For they considered all these habits[9] in that they are taught, that is, in that they are conveyed to students by an instructor having them; and so if anyone teaches others natural science based on the habit that he himself has, [that] natural science is called a teaching. And so if an instructor having the habit of logic teaches others the way itself of proceeding, that is, orders and methods, then the orders and methods are called teachings because they are conveyed to students by an instructor. Therefore in the definitions of order and method, those others accept teaching as the genus.

In no way° can I approve this position, since a consideration of 11 this type is completely accidental and particular. They act as if, when putting together a treatment on man's nature, they would have considered man as walking or speaking or dressed in white clothing and would have accepted accidents of this type in the definition of man. Of course, it is the nature of method and [the nature] of order that they are habits in man's soul that serve us in the conveying of the sciences and arts or in getting hold of knowledge altogether. But that someone endowed with the habit of logic should teach others logic and convey to others the same habits is accidental, alien to the nature of logic itself. Just as it happens with the art of building, that he who has the habit may teach it to others. The goal of his art is not this, however, but rather to effect

domum efficere. Potest enim eam artem aliquis habere, qui nullum alium doceat. Eadem ratione potest logicus habere hos instrumentales habitus et nemini eos tradere sed ipse per eos scientiam rerum investigare, vel aliis scientias tradere. Aliud enim est logicam docere, aliud est scientias et artes docere per habitum logicae. Genus igitur methodi non aliud sumendum est, quàm habitus mentis, differentia verò, instrumentalis, finalis autem causa cognitio rerum sive ab eodem acquirenda sive aliis tradenda.

12   Haec fuit absque dubio mens Galeni in principio Artis medicinalis, quando dixit, 'Tres sunt doctrinae, quae ordini innituntur,' de ordinibus enim loquens significavit tres esse modos ordinatè docendi scientias omnes et artes, sed non consideravit quòd ordines dicantur doctrinae, quia praeceptor logicus tres ordines doceat discipulos, hoc enim si respexisset, ipse quoque reprehensione non vacaret ob eam, quam diximus, rationem. Similiter Aristoteles in principio primi libri Posteriorum Analyticorum dicens omnem doctrinam ex praecedente cognitione fieri, nomine doctrinae non ipsius logicae, sed scientiarum traditionem intellexit. Illud igitur semper animadvertere debemus dum hos habitus vocamus doctrinas, ut nomen doctrinae non ad logicam docentem referamus, sed ad ipsam in usu positam, nempè ut intelligamus aliarum disciplinarum doctrinam per habitum logicae tanquàm per instrumentum factam, hic enim est logicae finis.

13   Haec autem, quae modò diximus, hoc argumento demonstrari possunt, quicquid alicui rei essentiale est et in eius definitione accipitur sive ut genus sive ut differentia, id ab ea re est prorsus inseparabile, ut non potest esse homo qui non sit animal et qui non sit rationalis. Sed potest esse tum ordo tum methodus tanquàm habitus in animo alicuius, quos ille neminem unquàm doceat,

a house. And someone who teaches no one else can still have this art. In the same way, a logician can have these instrumental habits and convey them to no one, but himself investigate by means of them the science of things; or he can convey the sciences on to others. For it is one thing to teach logic, another to teach the sciences and arts by means of the habit of logic. The genus of method therefore has to be taken as none other than *habit of mind*; and then the differentia, *instrumental*; and then the final cause, *knowledge of things*, whether to be acquired by the same person or to be conveyed to others.

Without doubt, this was what Galen had in mind at the begin- 12
ning of *The Art of Medicine*, when he said, "There are three [ways of] teaching, depending on order."[10] And when speaking about orders, he indicated that there are three orderly ways of teaching all the sciences and arts, but he did not consider orders to be called teachings just because the preceptor of logic teaches students three orders. And if he had regarded it this way, he too would not be free from censure, for the reason we said. Similarly, Aristotle, in the beginning of the first book of the *Posterior Analytics*,[11] saying that all teaching comes from preceding knowledge, understood by the name of teaching not the conveying of logic itself but of the sciences. Whenever therefore we call these habits teachings, we always ought to note that we refer the name of teaching not to teaching logic but to the same thing [i.e., to logic] put to use, that is, that we understand the teaching of other disciplines by means of the habit of logic, as made by means of an instrument. For that is the end of logic.

Now these things that we just said can be demonstrated by this 13
argument: Whatever is essential to some thing and is accepted in its definition, either as genus or as differentia, is utterly inseparable from that thing, just as there cannot be a man that is not animal and that is not rational. But there can be both order and method as habits in the soul of a man who never teaches them to

datur igitur ordo ac methodus, quibus nomen doctrinae ita acceptum non convenit. Ergo doctrina neque est eorum genus neque
differentia, proinde in eorum definitionibus secundùm eam significationem accipienda non est.

14   Semper tamen sunt instrumenta seu habitus mentis instrumentales, quia licèt in voce vel in scriptura ordo vel methodus sumi
possit, nihilominus ab habitu derivatur et habitum significat, quia
voces et literae sunt signa conceptuum, quos in animo habemus.

15   Haec de methodo amplè sumpta sufficiant, nunc ad eius species
transeamus.

: III :

De divisione Methodi, latè acceptae.

1   Dividitur methodus ita latè accepta in ordinem et methodum propriè dictam et nomine sui generis appellatam, eaque satis trita est
et vulgata divisio, quam unà cum differentiis dividentibus, quibus
alii utuntur, accipiendam et approbandam esse censemus. Hanc
igitur breviter et sine ulla disputatione referemus.

2   Cùm methodus sit habitus logicus docens procedere ab hoc ad
illud cognitionis adipiscendae gratia, duplex esse potest iste processus, aut enim res ipsas tractandas disponimus, ut prius de hac,
postea de illa agamus; aut à cognitione huius ducimur in cognitionem illius. Aliud enim est hanc rem prius esse cognoscendam,
quàm illam, aliud est ex hac re nota nos duci in cognitionem illius
ignotae. Hoc quidem methodi propriè sumptae munus est, illud
autem ordinis. Ordo enim nullam facit illationem huius rei ex illa,

anyone. There can, therefore, be order and method to which the name of teaching, thus accepted, is not appropriate. Therefore teaching is neither their genus nor their differentia. So according to this signification, it should not be accepted in their definitions.

Nevertheless, they are always instruments or instrumental hab- 14 its of [the] mind, because, granted that order or method can be taken in spoken or written words, nevertheless each is derived from habit and signifies habit, because spoken words and writings are signs of concepts that we have in our soul.

These [preliminaries] set down about [the genus of] method 15 are amply sufficient. Let us now move on to its species.

: III :

About the division of Method accepted in a broad sense.

Method, accepted in a broad sense, is divided into order and 1 method properly said and called by the same name as its genus. And we deem that this division, together with the differentiae that others use for dividing, is familiar and commonplace enough that it has to be accepted and approved. We will, therefore, recount it briefly and without any debate.

Since method is a logical habit teaching [someone] to proceed 2 from this to that for the sake of obtaining knowledge, such a pro- cedure can be of two types. For either we dispose the things to be treated so as to deal with this first, and afterward that, or we are led[12] from knowledge of this into knowledge of that. In the first, this thing has to be known prior to that. In the second, we are led from this known thing into knowledge of that unknown [thing]. The latter, of course, is the job of method taken properly; the former of order. For order makes no inference of this thing from

sed solùm disponit ea, quae tractanda sunt, ut si dicamus ordinem doctrinae postulare ut prius de coelo, quàm de elementis agamus. Methodus verò non disponit scientiae partes, sed à noto ducit nos in cognitionem ignoti inferens hoc ex illo, veluti quando à mutatione substantiae ducimur in cognitionem primae materiae; et quando ex aeterno motu venimus in cognitionem aeterni motoris immobilis vel ex aeterno motore immobili demonstramus aeternum motum. Propterea dicunt proprium esse ordinis disponere, methodi autem notificare.

3   Hoc discrimine methodus propriè accepta ab ordine dissidet, id autem fiet manifestius postquàm de utrisque diligenter locuti fuerimus.

4   De ordine quidem prius dicendum est, postea verò de methodo, nam author etiam quilibet scientiam aliquam vel artem scripturus, seu traditurus considerat ante omnia quo ordine eius disciplinae partes disponendae sint, postea in singula parte methodum quaerit, quae ex notis ducat in cognitionem eorum, quae ignorantur et quaeruntur; veluti prius decernit de animali generaliter agendum esse, mox de speciebus singulis. Postea verò de animali communi tractationem aggrediens methodos quaerit, quae ad animalis naturam, si latuerit, et ad eiusdem accidentia cognoscenda nos ducat; similiter cùm de homine et singulis aliis speciebus est locuturus.

5   Videtur etiam ordo universalius quiddam esse et latius extendi, quàm methodus; nam in ordine scientiam universam respicimus et eius partes inter se conferimus; methodus verò in unius rei quaesitae investigatione consistit sine ulla partium scientiae inter se comparatione. Quare omnino de ordine prius, quàm de methodo disserendum est; et primùm de ordine universè[8] sumpto.[9]

that, but only disposes the things that have to be treated, just as if
we said that the order of teaching demands that we deal with the
heavens prior to the elements. Method, however, does not dispose
the parts of a science, but leads us from the known into knowledge
of the unknown, inferring the latter from the former, just as when
from change of substance we are led into knowledge of first matter
and when from eternal motion we come into knowledge of the immobile eternal mover or from the immobile eternal mover we
demonstrate eternal motion.[13] Therefore we say that to dispose is
proper to order, and to make known [is proper] to method.

By this discriminating difference, method accepted properly is   3
distinguished from order.[14] This will become more manifest after
we have spoken carefully about each.

Order, of course, has to be talked about first, and then method   4
afterward. Now any author about to write down or convey some
science or art considers before all else in what order the parts of
his discipline have to be disposed. Afterward he inquires after the
method in each part that leads from the knowns into knowledge of
those things that are unknown and are inquired after, just as he
decides to deal first with animal in general and next with each species. But then afterward, undertaking a treatment on animal in
general, he inquires after the methods that lead us to the nature of
animal, if it has been hidden, and to its accidents that have [yet]
to be known; and similarly when he will speak about man and all
of the other species.

It appears then that order is something more universal and ex-   5
tends more broadly than does method, for in order, we regard a
science as a universal whole and compare its parts to each other,
whereas method consists in the investigation of one thing inquired
after without comparison of any of the science's parts to each
other. And so order has to be discussed altogether prior to method
—and first, order taken universally.

: IV :

In quo definitio ordinis ab aliis tradita et
difficultas contra eam exoriens exponitur.

1 Videntur omnes, qui de ordine locuti sunt, in hac ordinis definitione ferè consentire, Ordo doctrinae est habitus instrumentalis
seu instrumentum intellectuale, quo docemur cuiusque disciplinae
partes convenienter disponere.

2 De hac tamen definitione ego semper dubitavi, etsi enim falsa
esse non videtur, manca tamen et imperfecta est, cùm in ea animus
non conquiescat, sed dubius maneat et aliud quidpiam desideret,
quod in ipsa non exprimitur. In ea enim dicitur ordinem esse, quo
partes scientiae convenienter disponuntur, sed non exprimitur,
undenam ratio sumatur ita scientiam aliquam disponendi, ut bene
et convenienter dicatur esse disposita; proinde quaenam sit ista
conveniens dispositio nondum cognoscimus. Non est enim dicendum eam dispositionem à nobis temerè et absque ulla ratione et
penitus arbitratu nostro fieri, ut quemcumque ordinem servantes
et quocumque modo partes disciplinae disponentes vel transpo-
nentes dicamur[10] ordinem servare. Ita enim fieret ut ordo non esset
ordo, sed omni artificio° et utilitate carens inordinatio.

3 Propterea non rectam esse eorum sententiam arbitramur, qui
putant quamlibet scientiam et artem posse à nobis quocumque
voluerimus ordine tradi et explicari,[11] veluti artem medicam arbitrio nostro aequè scribi posse ordine compositivo, resolutivo atque
definitivo. De quorum errore postea uberius loquemur et eius rationem adducemus, cùm particulatim de singulis ordinibus sermonem faciemus.

: IV :

In which are laid out the definition of order conveyed by others and the problem arising with it.

Everyone who has spoken on order appears to agree, for the most 1
part, with the following definition: order of teaching is an instrumental habit or instrument of [the] understanding by which we
are taught to dispose appropriately the parts of each discipline.

But I have always had doubts about this definition. For al- 2
though it does not appear to be false, it is nevertheless deficient
and imperfect — our soul does not rest in it but remains doubtful
and desires something else that is not expressed in it. For in it,
order is said to be that in which the parts of a science are appropriately disposed; but from what one would get the rational way of
disposing the science so that it may be called well and appropriately disposed is not expressed; and so we do not yet know what
disposition is appropriate. For it should not be said that the disposition can be made by us rashly and without any reasoning and
completely at our discretion, so that, preserving whatever order
and disposing or transposing the parts of a discipline in whatever
way, we would be said to be preserving order. It would then happen that order would not be order but disorder, lacking all artful-
ness° and utility.

We therefore think to be incorrect the position of those who 3
hold that any science and any art can be conveyed and explicated
by us using whatever order we want, as if the medical art could be
written either using compositive, resolutive, or definitive order, as
we choose. We will speak below about their error more fully and
will adduce its reason when we discourse about each order one
by one.

4   Nunc satis est si dicamus neminem sanae mentis inficiari posse dandam esse necessariò aliquam certam rationem seu certam normam, à qua semper sumatur haec dispositionis rectitudo et convenientia, ut haec dispositio dicatur conveniens, illa verò non conveniens. Hanc autem normam unam esse necesse est, nisi enim una esset omnis ordinis ratio, ex qua ordo quilibet dicatur esse conveniens, non esset ordo vox univoca, sed ambigua, de qua ut de re una sermo haberi non posset. Quo fieret ut omnes aliorum de ordinibus tractationes vitiosae essent, quia omnes in ambiguo laborassent.

: V :

Quid alii ad hanc quaestionem dicere videantur.

1   Difficultas haec ab iis, qui de hac re scripserunt, non admodum fuit animadversa, vel enim de ea nihil dixerunt vel si quid ex eorum dictis colligi potest, id ab eis inconsideratè prolatum fuit. Quicquid enim statuerunt esse rationem rectè ordinandi omnes disciplinas, id tamen in definitione ordinis non expresserunt, cùm omnino fuerit exprimendum, ut ipsius ordinis natura benè declaretur.

2   Aliqui videntur existimare eam esse huius convenientiae rationem, ut omnia, quae in qualibet scientia considerantur, ab uno aliquo pendeant et ad illud referantur, quod cùm esse possit vel unum principium vel unum medium vel unus finis, hinc exoriri dicunt tres ordinis species, compositivum, resolutivum, definitivum.

For now it is enough if we say that no one of sound mind can   4
deny that there necessarily has to be some definite reason or definite norm from which the correctness and appropriateness of the
disposition may always be taken, so that this disposition may be
called appropriate and that one not appropriate. Moreover it is
necessary that there be one such norm, for unless there were, for
every order, one reason by which any order may be said to be appropriate, then "order" would be an ambiguous and not a univocal
word. It would not be possible to discourse about it as about one
thing. And by this it would happen that all treatments on order
by others would be flawed because all would have labored in ambiguity.

: V :

What others appear to say about this question.

This problem was not noted at all by those who wrote on this is-   1
sue. Either they wrote nothing on it or, if anything can be gathered from their writings, it was advanced by them without due
consideration. For whatever they decided the rational way of correctly ordering all disciplines was, they nevertheless did not express it in the definition of order, although, for the nature of order
itself to be well clarified, this would have had to be completely
explicit.

Some appear to judge that the way to think about this appro-   2
priateness [of order] is that all things that are considered in any
science depend on some one thing and are referred to it. Since this
could be either one beginning, one middle, or one end, they then
say that three species of order arise — compositive, resolutive, and
definitive.

3   Sententiam hanc diligenter expendere non possumus, nisi de speciebus ordinis loquamur, idcirco ad alium locum eam considerandam remittemus, ubi ostendemus pendentiam hanc ab uno neque ordinis naturam declarare neque omni ordini competere.

4   Multi videntur illi sententiae adhaerere, quam fusè aliqui philosophi explicarunt, dum praefationem primi libri Physicorum Aristotelis interpretarentur. Ibi namque Averrois opinionem refellunt, qui dicit Aristotelem proponere in naturali philosophia ordinem servandum esse ab universalibus ad particularia, propterea quòd universale est nobis notius, quàm particularia, et à notioribus nobis semper est progrediendum.

5   At dicunt hi deceptum fuisse Averroem, qui id attribuit ordini, quod potius methodo ac viae tribuendum est, proprium enim est viae seu methodi ut à notis progrediatur ad ignota, quia finis methodi est facere ex notis cognitionem eius, quod ignoratur. Ordo autem non requirit ut à notioribus auspicemur, sed solùm ut ea anteponantur, quae natura priora sunt.

6   Horum igitur sententiam si sequamur, ratio ac norma bene et convenienter omnem disciplinam disponendi non alia est, quàm ipse naturalis ordo rerum considerandarum, ut illa sit partium alicuius scientiae conveniens dispositio, quae rerum in ea tractatarum naturalem ordinem imitetur[12] et sequatur, ut cùm ordine naturae elementa praecedant mistum, ordo doctrinae rectus et conveniens ille est, quo prius de elementis agitur, quàm de mistis; quòd si prius de mistis ageretur, prava esset dispositio; quia ordini ipsarum rerum naturali minimè esset consentanea.

7   Sententiam hanc testimonio Aristotelis confirmant, qui scientiam naturalem auspicatus est à tractatione de primis principiis,

We cannot weigh this position carefully unless we speak about   3
the species of order. Therefore we will leave it to be considered in
another place, where we will show that this dependence on one
thing neither makes the nature of order clear nor appertains to every order.

Many appear to stick to that position which some philosophers   4
explicated at length when they were commenting on the preface to
the first book of Aristotle's *Physics*.[15] But there they refute the
opinion of Averroës, who says that Aristotle sets out that, in natural philosophy, an order from universals to particulars has to be
maintained because the universal is more known to us than particulars and progression always has to be from the things more
known to us.

They say that Averroës was deceived and that he attributed to   5
order that which is better ascribed to method and way. For it is
proper to way or to method to progress from knowns to un-
knowns, because the end of method is to bring about from the
knowns, knowledge of that which is unknown. Order, however,
does not require that we commence with things more known, but
only that those things be placed first that are by nature prior.

If therefore we follow their position, the way and the norm for   6
disposing every discipline well and appropriately is nothing other
than the natural order itself of the things being considered, so that
the appropriate disposition of the parts of some science is that
which imitates and follows in the natural order of the things
treated, just as, since in the natural order the elements precede
mixed [body], the correct and appropriate order for teaching is
that in which the elements are dealt with prior to mixed [bodies];
if mixed [bodies] were dealt with first the disposition would be
defective, because it would not agree at all with the natural order
of the things themselves.

They confirm this position with the testimony of Aristotle, who   7
commenced natural science with a treatment on first beginning-

quae tamen occulta et incognita erant et cognitu maximè difficilia, ipsisque composita multò notiora erant. Non possumus igitur dicere Aristotelem prius egisse de principiis, quòd notiora fuerint, id enim est manifestè falsum. Itaque confiteri cogimur ipsum de principiis agere primùm voluisse ea tantùm ratione ductum quia principia sunt, principia enim quatenus sunt principia, natura priora sunt compositis et effectis. Ratio igitur ordinandi omnes disciplinas videtur esse ipse naturalis ordo rerum considerandarum.

: VI :

In quo dicta sententia refellitur et ostenditur
rationem ordinandi non ab ipsa rerum
natura, sed à nostra cognitione desumi.

1   Ego verò existimo sententiam hanc admittendam non esse, sed veram et firmis nixam fundamentis esse opinionem Averrois, quam illi reiecerunt. Revera enim non ex ipsa rerum considerandarum natura sumitur ratio ordinandi scientias et disciplinas omnes, sed ex meliore ac faciliore nostra cognitione. Non enim scientiam aliquam hoc potius, quàm illo modo disponimus, quòd hic sit rerum considerandarum naturalis ordo prout extra animum sunt; sed quia ita melius et facilius ab omnibus ea scientia discetur.

2   Quam quidem veritatem comprobare et contrariam opinionem evertere non erit difficile tum ratione duce tum multis Aristotelis testimoniis.

principles that were nevertheless hidden and unknown and very difficult to know and compared to which composites were much more known. We cannot say, therefore, that Aristotle dealt first with beginning-principles because they were more known, for that is manifestly false. And so we are forced to confess that he wanted to deal with beginning-principles first and was led [to do so] only for the reason that they are beginning-principles. For beginning-principles, insofar as they are beginning-principles, are by nature prior to composites and effects. It appears, therefore, that the rational way of ordering all disciplines is the natural order itself of the things being considered.

: VI :

In which the said position is refuted and it is shown that the rational way of ordering is drawn not from the nature itself of things, but from our knowledge.

But I judge that this position does not have to be admitted. Instead, the opinion of Averroës that they rejected is true and relies on firm foundations. For in truth, the rational way of ordering the sciences and all the disciplines is not taken from the nature itself of the things being considered but from our better and easier knowledge. For we dispose some science in one way rather than another, not because this is the natural order of the things being considered as they are outside man's soul, but because in that way the science can be learned better and more easily by all.

Of course, it will not be difficult to confirm this truth and overturn the contrary opinion, both by the guide of reason and by multiple testimonies of Aristotle.

1

2

3   Primùm quidem, si admitteremus convenientem ordinem in singula disciplina desumi ab ordine naturali rerum consideratarum, sequeretur nullum dari alium ordinem, quàm compositivum, consequentia manifesta est, in omnibus enim simplicia et principia sunt natura priora compositis et effectis, erit igitur semper à simplicibus inchoandum et ab iis ad composita progrediendum in omnibus disciplinis. Unus igitur dabitur ordo compositivus. Sed hoc consequens falsum est et ab Aristotele et à Galeno et à communi omnium consensu reprobatum, Galenus enim tres ordines statuit, compositivum, resolutivum et definitivum, quam sententiam alii omnes, qui posterioribus temporibus de hac re scripserunt, secuti sunt. Aristoteles autem de ordine resolutivo locutus est in contextu 23 septimi libri Metaphysicorum dicens artes omnes effectrices tradendas esse ordine resolutivo, qui est à notione finis, ut si tradenda sit ars aedificatoria, primum eius artis discendae principium erit praecognitio finis, id est ipsa definitio domus aedificandae, à qua transitus fit ad fundamenta, parietes, tectum, et ab his tandem ad lapides, lateres, caementa et ligna, quae prima et simplicissima illius artis elementa sunt, in quibus desinit ars ipsa docens, et à quibus postea ars operans exorditur.

4   Idem videtur significare in capite octavo septimi libri de Moribus, quando inquit finem in rebus agendis principium esse, quemadmodum in mathematicis suppositiones. In omnibus enim disciplinis, quae ad finem aliquem tendunt, finis ipse, licèt reipsa postremus sit, est tamen totius cognitionis principium. Significavit autem iis verbis Aristoteles morales disciplinas ordine resolutivo esse tradendas, quo quidem ordine qui non videt Aristotelem scripsisse libros de Moribus, caecus est, nam à felicitate, quae est

First, of course, if we admitted that the appropriate order in 3
each discipline is drawn from the natural order of the things under
consideration, it would follow that there is no order other than
compositive.[16] The consequence is manifest. For in everything,
beginning-principles and simple things are by nature prior to effects and composite things. In all disciplines, therefore, one always
has to start with simples and progress from these to composites.
There would, therefore, be one order, the compositive. But this
consequence is false and renounced by Aristotle, by Galen, and
by the common consensus of all. For Galen thought that there
are three orders—compositive, resolutive, and definitive—and this
position was followed by all others who in later times wrote on the
issue. Moreover, Aristotle spoke on resolutive order in text no. 23
of the seventh book of the *Metaphysics*,[17] saying all effective arts
should be conveyed using resolutive order, which is from a notion
of the end. Just as, if the art of building is to be conveyed, the first
beginning-principle of the art to be learned will be the prior
knowledge of the end, that is, the definition of the house to be
built. From this, passage is made to the foundation, walls, roof,
and from these finally to stones, tiles, mortar, and wood; these
[last] are the first and simplest elements of the art. In them the art
as taught ends, and from them the art as practiced afterward
begins.

He appears to have indicated the same in the eighth chapter of 4
the seventh book of the *Ethics*,[18] when he says that in doing things
the end is the beginning-principle, just as in mathematics suppositions are. For in all disciplines that aim toward some end, the end
itself, even granted that it be in reality what is final, is nevertheless
the beginning-principle of the whole [body of] knowledge. And so
with these words, Aristotle indicated that moral disciplines have
to be conveyed using resolutive order. Of course, anyone who does
not see that Aristotle wrote the *Ethics* books using this order is
blind. For he commenced with felicity, which is the ultimate end

ultimus rerum agendarum finis, auspicatus est, de qua in toto primo libro pertractavit, mox in secundo ad virtutem transivit, quae secundùm naturam prior est felicitate, est enim felicitas apud Aristotelem actio ex virtute, non virtus ipsa.

5   Libros quoque Analyticos tum Priores tum Posteriores ordine resolutivo scriptos esse manifestum est, in illis namque à definitione syllogismi, in his à definitione scientiae ac demonstrationis exordium sumitur, et ab his ad principiorum investigationem proceditur.

6   Patet igitur scriptas esse ab Aristotele multas disciplinas ordine resolutivo, attamen hic ordo non sequitur ordinem rerum naturalem, imò est ei contrarius, ut apertè asserit Aristoteles in illo 23 contextu septimi Metaphysicorum. Ergo falsum est id, quod illi dicunt, rationem ordinandi omnes disciplinas sumi ab ordine naturali rerum considerandarum.

7   Praeterea si aliquam scientiam scribendam nobis proponamus, ut scientiam naturalem, dato, nec tamen concesso quòd eius partes non ex nostra cognitione, sed ex ordine rerum naturali disponendae sint, adhuc perplexi sumus, adhuc non cognoscimus quo ordine scientiam naturalem scribere debeamus, quoniam ordo naturae non unus est, sed multiplex vel saltem duplex, non solùm enim simplicia sunt natura priora compositis, sed etiam composita simplicibus, ut ex Aristotele colligimus in primo capite secundi libri de Partibus animalium. Si namque ordinem naturae intelligamus habita ratione° generationis naturae, elementa sunt priora misto, mistum animali, animal homine. Quòd si ordinem naturae intelligamus ordinem perfectionis seu ordinem scopi et intentionis naturae, homo est natura prior animali, animal misto, mistum elementis. Cùm igitur utroque contrario ordine utentes dicamus[13] ordinem naturae servare, non possumus ab ipsa rerum natura

of doing things. He dealt with it through the whole first book. Next, in the second book, he passed on to virtue, which, according to nature, is prior to felicity. For felicity in Aristotle is acting out of virtue[19] and is not virtue itself.

It is also manifest that both the *Prior* and *Posterior* [*Analytics*] are 5 written using resolutive order. For in the first, the beginning is taken from the definition of syllogism, in the second from the definition of scientific knowledge and demonstration. And from these proceeds the investigation of beginning-principles.

It is patent, therefore, that Aristotle wrote about many disciplines using resolutive order. But this order does not follow the natural order of things. Rather, it is contrary to it, as Aristotle plainly asserts in text no. 23 of the seventh [book] of the *Metaphysics*.[20] Therefore what they say is false, that the rational way of ordering all disciplines is taken from the natural order of the things being considered.

And moreover, if we set out to write about some science, such 7 as natural science, having presumed, though not yet conceded, that its parts are disposed not from our knowledge but from the natural order of things, we are still perplexed. We still do not know in what order we ought to write about natural science, since the order of nature is not one but of multiple or at least of two types. For not only are simple things by nature prior to composites, but composites are [prior] to simples also, as we gather from Aristotle in the first chapter of the second book of *On the Parts of Animals*.[21] For if we are to understand the order of nature by taking account° of the generation of nature, then elements are prior to mixed [body], mixed [body] to animal, and animal to man. But if we are to understand the order of nature as the order of perfection or the order of the goal and intention of nature, then man is by nature prior to animal, animal to mixed [body], mixed [body] to elements. Since we can, therefore, say that those using either of the contrary orders maintains the order of nature, we cannot from

certum ordinem desumere, quo scientiam naturalem disponamus. Vel ad id confugiemus ut dicamus naturalem scientiam posse arbitratu nostro utrovis ordine tradi, quod tamen falsum est et Aristoteli adversatur, qui in prooemio primi libri Physicorum ostendit non posse res naturales benè cognosci, nisi ex primorum principiorum cognitione. Horum igitur sententia ambiguitatem habet, dum dicunt rationem ordinandi omnes disciplinas sumendam semper esse ex ordine ipsarum rerum naturali.

8   Possumus etiam eos proprio ipsorum argumento ad veritatis confessionem compellere, si prius ex Aristotele in Posterioribus Analyticis sumamus illam esse veram cuiusque rei rationem et causam, quam habentes desinimus quaerere amplius cur sit; contra verò illam non esse veram rei causam, quam habentes adhuc quaerimus cur sit. Cùm itaque Aristoteles in prooemio primi libri Physicorum dicat se in tractatione rerum naturalium velle à primis principiis auspicari, si nos rationem percontemur quare à primis principiis exordiri velit, ipsi certè respondebunt, quia vult servare ordinem naturae. At hanc non esse veram rationem eius, quod quaeritur, manifestum est, propterea quòd in hac responsione animus non quiescit, adhuc enim quaerimus, quare vult ordinem naturae servare? Ad quam interrogationem ipsi, velint nolint, coguntur respondere, ut optimam rerum naturalium cognitionem consequamur. Adeptio igitur perfectae scientiae est prima causa et vera ratio illius ordinis, quae si primo loco adducta fuisset, ita satisfecisset quaestioni cur sit, ut non opus fuisset amplius eandem quaestionem repetere. À principiis enim exordiendum fuit, quoniam ex principiorum cognitione perfectam assequemur scientiam rerum naturalium. Haec enim revera est causa finalis omnis ordinis, in qua animus noster absque ullo dubio conquiescit. Et hanc

the very nature of things draw a definite order in which we can dispose natural science. Or we might resort to saying that natural science can be conveyed using either order, at our discretion. But this is false and is opposed by Aristotle, who shows in the proem of the first book of the *Physics*[22] that natural things cannot be known well except from knowledge of first beginning-principles. When they say, therefore, that the rational way of ordering all disciplines always has to be taken from the natural order of things themselves, their position is ambiguous.

Now we can compel them into confession of the truth by means of their own argument, if we first take the following from Aristotle's *Posterior Analytics*:[23] That is the true reason and cause of some thing if, having it, we cease to inquire further why the thing is; and on the other hand, that is not the true cause, if, having it, we still inquire why the thing is. Now Aristotle says in the proem of the first book of the *Physics*[24] that in a treatment of natural things, he wants to commence with first beginning-principles. If we probe for the reason why he wants to begin with first beginning-principles, they will certainly respond that it is because he wants to maintain the order of nature. But it is manifest that this is not the true reason for what was inquired after, because in this response our soul does not rest. For we further inquire: why does he want to maintain the order of nature? To this question, they are forced to respond, whether they want to be or not, that it is so that we may gain optimal knowledge of natural things. Obtaining perfect scientific knowledge, therefore, is the first cause and true reason for that order. And had that been adduced in the first place, it would have answered the question, "Why is it?," so that the same question would not need repeating. He had to begin with beginning-principles, because from knowledge of beginning-principles we will secure perfect scientific knowledge of natural things. And this, in truth, is the final cause of every order; in it our soul rests without any doubt. And this is the very [cause and

8

35

ipsam affert ibi Aristoteles, tractationis enim de primis principiis non hanc rationem affert, quam illi dicunt, nempè quòd sint principia et secundùm naturam prima, sed quia perfecta rerum naturalium cognitio pendet ex primorum principiorum cognitione. Nullo igitur alio medio° ostendit eius ordinis rationem, quàm nostra cognitione, hanc vult ibi Aristoteles esse rationem ordinandi omnes disciplinas, quare locus ille apertissimè favet nostrae opinioni.

9 Sed res haec magis est manifesta, quàm ut pluribus argumentis indigeat, traduntur enim disciplinae non ut in rebus ipsis ordo statuatur, eum enim iam natura ipsa constituit; sed ut nos discamus. Eo igitur ordine utimur, quo melius ac facilius discamus, haec est vera ratio ordinandi, ad quam et res ipsa nos ducit et ipsa quoque vocum significatio, appellatur enim à cunctis ordo doctrinae, non ordo naturae.

10 Sed causa, ut puto, erroris multorum fuit, quoniam quandoque haec duo simul sunt, contingit enim ut eadem scientia et naturae ordinem servet et ordinem melioris cognitionis, ideò opinantur talem servari ordinem ea ratione, qua est ordo naturae, non quatenus per eum melior cognitio habenda est, quemadmodum qui putant aliquam figuram habere tres angulos aequales duobus rectis quatenus est aequilatera, non quatenus est triangularis, quoniam contingit eandem figuram triangulum esse et aequilaterum. Quando igitur evenit ut in scientia ordo naturae generantis servetur, id non fit quatenus est ordo naturae, sed quatenus est ordo melioris nostrae cognitionis, unde etiam appellationem suscepit, vocatur enim ab omnibus, quemadmodum diximus, ordo doctrinae.

reason] Aristotle brings forward there. The reason he brings forward for a treatment on first beginning-principles is not what others say, namely that they are beginning-principles and first according to nature, but that perfect knowledge of natural things depends on knowledge of first beginning-principles. Aristotle shows, therefore, that the reason for his order has no other purpose° than our knowledge, and there he wants this to be the way of ordering all disciplines. Thus this passage very plainly favors our opinion.

Now this is too manifest to be in need of more arguments. For disciplines are conveyed not so that the order may be fixed in the things themselves — for nature itself has already established that — but so that we may learn. We use, therefore, that order by which we may learn better and more easily. This is the true and rational way of ordering; both the thing itself and also the very signification of the word lead us to this. For it is called by everyone the order of teaching, not the order of nature. 9

Now the cause of error by many was, I hold, that sometimes the two occur together. For it happens that the same science may maintain both the order of nature and the order of better knowledge. Because of this, some were of the opinion that such order is maintained for the reason that it is the order of nature and not insofar as, by means of it, better knowledge will be had, just as some hold that a given figure has three angles equal in total magnitude to two right angles, insofar as it is equilateral and not insofar as it is triangular, when it happens that the same figure is a triangle and an equilateral. When it happens, therefore, that in a science the order of generating nature is maintained, this happens not insofar as it is the order of nature, but insofar as it is the order of our better knowledge. It was surely from this that it took on its appellation. For it is called by everyone, just as we said, the order of teaching. 10

: VII :

*In quo ex quorumdam apud Aristotelem locorum
observatione idem demonstratur.*

1   Horum, quae diximus, efficacissima confirmatio sumi potest ex
plurium locorum observatione, in quibus Aristoteles ordine naturae neglecto voluit ordinem nostrae melioris cognitionis servare.
Quemadmodum enim quando ordo cognitionis est etiam naturae
ordo, magna datur occasio errandi, cùm non appareat utrum duorum author ipse respexerit; ita quando disiuncti atque etiam contrarii sunt, nempè si contingat ordinem naturae non esse ordinem
melioris cognitionis, tota rei veritas clara nobis et perspicua redditur, dum videmus sperni ordinem naturae et solum ordinem cognitionis servari.

2   Protulit hanc sententiam Aristoteles, eamque in suis libris executus est. Nam in principio quinti libri Metaphysicorum clara voce
testatur non semper in disciplinis principium cognitionis esse
etiam principium rei, sed id, unde facilius discamus.

3   In primo quoque libro de Moribus capite quarto inquit dubium
esse an ea moralis disciplina tradenda sit ordine compositivo,
quem vocat à principiis, an resolutivo, quem vocat ad principia,
qua de re ait Platonem rectè dubitasse. Postea quaestionem dissolvens dicit nil aliud esse respiciendum nisi ut à notis incipiamus,
non considerando an sint principia rei, necne. Quare iubet ordinem cognitionis semper servari, non semper ordinem naturae.

: VII :

In which the same is demonstrated from observation of some passages in Aristotle.

The most effectual confirmation of the things we have said can be 1
taken from observation of many passages in which Aristotle, having neglected the order of nature, wanted to maintain the order of
our better knowledge. Now just as when the order of knowledge is
also the order of nature, there is a great opportunity for erring,
because it is not apparent which of the two the author himself was
regarding, so also when the two are disjunct or even contrary —
that is, if it happens that the order of nature is not the order of
better knowledge — the whole truth of the issue is rendered clear
and plain to us when we see that the order of nature is left aside
and only the order of knowledge is maintained.

Aristotle advanced this position and followed it in his books. 2
For in the beginning of the fifth book of the *Metaphysics*[25] he attests in a clear voice that in disciplines the beginning (*principium*)
of knowledge is not always the beginning-principle (*principium*) of
the thing, but that from which we may learn more easily.

Also, in the fourth chapter of the first book of the *Ethics*,[26] he 3
says there is doubt whether the moral discipline should be conveyed using compositive order, which he calls "from beginning-
principles," or using resolutive, which he calls "to beginning-
principles." About the issue, he says, Plato correctly had doubts.
Afterward,[27] resolving the question, he says nothing else has to be
regarded except that we start from things known, not considering
whether they are or are not the beginning-principles of the thing.
He thus demands that the order of knowledge always be maintained, but not always the order of nature.

4   Miror autem quòd aliqui eum locum interpretantes dicant illa
verba ad viam doctrinae, non ad ordinem pertinere, quòd conditio
methodi sit, non ordinis, ut à notioribus nobis progrediamur, eo
fortasse argumento ducti quòd ibi Aristoteles non nominet ordinem, sed dicat ὁδὸν, quae via est. Quod quidem argumentum
nullius est roboris, cùm possit et ordo et methodus communi appellatione vocari ὁδὸς, quare vox quoque composita μέθοδος
saepe pro ordine accipitur, ut multis in locis apud Aristotelem et
alios Graecos authores observare possumus. Revera enim si ipsius
vocis significationem spectemus, non minùs ordinem significare
potest, quàm illam, quae propriè methodus dici solet, nam et haec
et ille via quaedam est et quidam progressus ex hoc ad illud. Distinctio autem vocabulorum non ex ipsa vocum proprietate de-
sumpta est, sed ad evitandam ambiguitatem et obscuritatem in-
venta° fuit. Ratio itaque illorum frivola est.

5   Quòd autem Aristoteles ibi de ordine loquatur, non de methodo, manifestum est si eius verba perpendamus, dicit enim dubium esse an à principiis an ad principia via tenenda sit. At certum
est, si de methodo haec intelligantur, significari his verbis duas
species demonstrationis, esset igitur quaestio Aristotelis utra demonstrationis specie uti debeat, an ea, quae est à priori, an ea,
quae à posteriori. Quae certè vana esset dubitatio et contraria iis,
quae antè in tertio capite dixerat Aristoteles,[14] ubi protestatus erat
se nulla usurum esse demonstratione, sed argumentis tantùm
communibus et vulgaribus, qualia materia subiecta requirit. Itaque
non potest in quarto capite dubitare utra demonstrationis specie
uti debeat. Idque rationi consentaneum est, quia si in tertio capite
de methodo verba fecerat, quam ad res ignotas declarandas in iis

Now I wonder at the fact that some, commenting on this pas- 4
sage, say that these words pertain to the way of teaching but not to
the order, since it is a characteristic of method, not of order, that
we progress from the things more known to us. Perhaps they are
led by the argument that Aristotle does not there name it "order,"
but says *hodos*, which is way. This argument, of course, is of no
weight, since both order and method can be called by the common
appellation *hodos*. Therefore also, the composite word, *methodos*, is
often accepted for order, as we can observe in many passages in
Aristotle and other Greek authors. In truth, if we look at the signification of the word itself, it can signify order no less than that
which is normally called method taken properly. For both the latter and the former are each some sort of way and some sort of
progression from this to that. Moreover, the distinction between
the terms was not drawn from any property itself of the words,
but was invented° to avoid ambiguity and obscurity. And so their
reasoning is frivolous.

Moreover, that Aristotle there speaks about order and not 5
about method is manifest, if we examine his words. For he says
there is a doubt whether the way should be taken either from
beginning-principles or to beginning-principles. But it is certain,
that if they are understood as being about method, two species of
demonstration are signified by these words. Aristotle's question
therefore would be which species of demonstration ought to be
used, that which is *a priori* or that which is *a posteriori*. Such
doubting would certainly be vain and contrary to what Aristotle
had said earlier in the third chapter,[28] where he had insisted he
would use no demonstration except general and commonplace arguments, such as the subject matter requires. And so in the fourth
chapter there can be no doubt which species of demonstration
ought to be used. Now this agrees with reason, because if in the
third chapter he had offered up some words on the method that
he will maintain in his books when making unknown things clear,

libris servaturus erat, postea in quarto capite non debuit iterum de methodo loqui, sed de ordine, ut significat verbum illud 'incipere' quod non viae doctrinae, sed soli ordini convenit ob eam rationem, quam in sequentibus declarabimus.

6   Plures etiam apud Aristotelem loci notari possunt, in quibus ipse ordine naturae praeterito ordinem facilioris doctrinae sectatus est, in libro enim Categoriarum relationem qualitati anteposuit propter faciliorem doctrinam, quia in capite de quantitate relationis mentionem fecerat, tamen ordine naturae qualitas anteponenda erat, quia absolutum° est natura prius respectivo.

7   Porphyrius quoque caput de specie anteposuit capiti de differentia facilioris doctrinae gratia, cùm tamen differentia sit natura prior specie.

8   Et Aristoteles in secundo libro de Anima, in tractatione de quinque externis sensibus à visu exordium sumpsit, ultimo autem loco de tactu locutus est; tamen ordine naturae tactus praecedere debuit, tanquàm omnium sensuum imperfectissimus et communissimus. Ordinem igitur naturae sprevit propter faciliorem doctrinam. Cùm enim in singulis sensibus cognoscenda fuerint obiecta et media et organa, ut cognosceremus operationem fieri ex actione obiecti in organum per medium transferens speciem° ab obiecto ad organum, animadvertit Aristoteles ita esse incognita et obscura in tactu organum et medium, ut si à tactu incepisset, iure dubitare potuissemus an quilibet sensus organum et medium habeat. Ideò à visu auspicari voluit, in quo ea omnia maximè dilucida et distincta erant, ut cùm in visu et aliis sensibus nobilioribus cognitum esset operationem fieri ex translatione speciei° ab obiecto ad organum per medium, illud idem in aliis quoque sensibus, in quibus obscurissimum erat, fateremur, videlicet in gustu et in tactu, quos ibi Aristoteles in postremo loco posuit. Esse autem difficilia cognitu organa et media horum duorum sensuum declarat ipsa

afterward, in the fourth book, he ought not to speak again on method, but on order, as indicated by the verb "start," which is appropriate not to the way of teaching but only to order, for [a] reason we will make clear in the following.

Now many passages in Aristotle can be noted in which, the 6 order of nature having been passed over, he followed the order of easier teaching. In the book on the *Categories*, he placed relation before quality on account of easier teaching, because he had made mention of relation in the chapter on quantity. But by the order of nature, quality had to have been placed first, because absolute° is by nature prior to relative.[29]

Porphyry too placed the chapter on species before the chapter 7 on differentia for the sake of easier teaching, while by nature differentia is prior to species.

And Aristotle in the second book of *On the Soul*,[30] in the treat- 8 ment on the five external senses, took his beginning from sight, and spoke about touch in the last passage, even though, as the most imperfect and common of all the senses, touch ought, by the order of nature, to have gone first. He therefore left aside the order of nature on account of easier teaching. Since the objects, media, and organs in each of the senses have to be known for us to know the activity occurring from the action of an object on the organ by means of the medium transferring the species-image° from the object to the organ, Aristotle noted that the organ and medium are so unknown and obscure with touch that if he had begun from touch, we could justly have doubted whether each sense had an organ and a medium. Because of this, he wanted to commence with sight, in which all these were especially lucid and distinct, so that—since it is known that in sight and the other more noble senses their activity occurs by transference of the species-image° from object to organ by means of a medium—we may acknowledge that it is also the same in the other senses in which it was most obscure, namely in taste and in touch, which Aristotle placed

43

Aristotelis tractatio, ex qua nihil certi de his desumi potest, cùm ipse nihil exprimat, quod in iis duobus sensibus organum aut medium dici possit, sed haec leviter tangat et sicco pede praetereat.

9   Haec fuit vera ratio illius ordinis, quo Aristoteles sensus ipsos disposuit,[15] aliis adhuc incognita, omnes enim unà cum Averroe decepti nullam aliam afferre solent eius ordinis rationem, quàm quòd Aristoteles servare voluit ordinem nobilitatis. Scilicet si Aristoteles à tactu exorsus esset, non minùs idoneam eius quoque[16] contrarii ordinis rationem allaturi erant, dixissent enim ipsum ordine naturae generantis uti voluisse, natura enim ab imperfectis ad perfecta progredi solet, dum generat. Igitur quemcumque voluit ordinem servare Aristoteles potuit, siquidem pro eius excusatione ratio aliqua illius ordinis nobis non erat defutura, quo nihil ineptius dici posse demonstrabimus in sequentibus.

10   At verò si eos percontaremur, quare in tractatione de sensibus voluit Aristoteles ordinem servare à nobiliore, in tractatione autem de partibus animae non servavit eundem ordinem, sed potius contrarium, qui est ab ignobiliore? Nullam huius diversitatis rationem, quam adducerent, haberent. Nos autem horum omnium rationem optimam afferimus dum dicimus ordinem utrobique servatum esse facilioris ac melioris doctrinae, haec enim postulavit ut à parte animae ignobiliore et à[17] nobiliore sensu tractationis exordium sumeretur.

11   Simile quiddam[18] pro huius veritatis confirmatione apud Aristotelem legere possumus in primo de Historia animalium capite sexto, et in secundo de Partibus animalium capite decimo, quibus

there in the final passage.[31] Moreover, Aristotle's treatment itself makes clear that the organs and media of these two senses are difficult to know. Nothing definite can be drawn about these from his treatment, for he himself does not say anything expressly that could be said to be about the organ or medium of these two senses; instead, he touches on them lightly and passes by without wading in.

This was the true reason for the order in which Aristotle disposed [his treatment of] these senses, hitherto unknown to others. For together with Averroës all were deceived. Usually they brought forward no other reason for his order than that Aristotle wanted to maintain the order of nobility. Of course, if Aristotle had begun with touch, they would have also brought forward a reason, no less fitting, for the contrary order. For they would have said that he wanted to use the order of nature generating; for when it generates, nature normally progresses from imperfect to perfect. Aristotle could, therefore, have maintained whatever order he wanted, since, to give him an excuse, we would not lack some reason for that order. In what follows we will demonstrate that nothing more inept than this could be said.

For truly if we probe for why, in his treatment on the senses, Aristotle wanted to maintain an order from the more noble, but in the treatment on the parts of the soul, he maintained not the same order but instead the contrary, that is, from the more ignoble, they could adduce no reason for this difference. We now bring forward the best reason for all this, when we say that in both places the order of easier and better teaching was maintained. For this demanded that the beginning of the treatment be taken from the more ignoble part of the soul and from the more noble sense.

For confirmation of the truth of this, we can read something similar in Aristotle in the sixth chapter of the first book of *On the History of Animals*,[32] and in the tenth chapter of the second book of *On the Parts of Animals*.[33] In these passages he sets out that the

in locis proponit dicendum prius de partibus hominis, quàm de partibus aliorum animalium, quia nobis, qui homines sumus, promptiores ac notiores sunt, ex quarum cognitione in notitiam° partium aliorum animalium facilius venire poterimus. Certum est autem Aristotelem ibi de ordine doctrinae loqui, non de via, ordinem tamen dicit esse servandum à notioribus, neque dicit ab homine auspicandum esse ut à nobiliore, sed ut à[19] notiore.

12   Libros quoque de animalibus anteposuit Aristoteles libris de plantis, ut ipse testatur in fine libelli de Longitudine et brevitate vitae, idque non ob aliam causam, quàm propter nostram faciliorem cognitionem, multa enim sunt in animalibus, quibus quaedam in stirpibus proportione respondent, quemadmodum radix, quae instar oris est, et alia eiusmodi, quae omnia facilius cognoscentur in plantis, si prius in animalibus cognita fuerint, in quibus distinctiora et evidentiora sunt.

∴ VIII ∴

In quo argumentum adversariorum solvitur
et ostenditur quomodo ordo doctrinae
sit à notioribus nobis.

1   Refutata eorum opinione, qui dicunt rationem ordinandi non sumi à nostra cognitione, sed à rerum tractandarum natura facile est ipsorum argumentum solvere, quod ex primo Physicorum libro sumebatur, dicimus enim eos falsum accipere cùm dicunt Aristotelem à principiorum tractatione auspicari voluisse non quatenus

parts of man have to be talked about prior to the parts of other animals, because to us, who are men, they are more accessible and more known. From knowledge of them we will more easily be able to come to knowledge° of the parts of other animals. Moreover it is certain that Aristotle speaks there about the order of teaching, not about the way. Nevertheless, he says the order from things more known has to be maintained. He says that it has to commence with man, not as from the more noble, but as from the more known.

Also, Aristotle placed the books on animals before the books on 12 plants, as he attests at the end of the little book *On Length and Shortness of Life*,[34] and this for no other cause than on account of our easier knowledge. For there are many things in animals to which some things in plants correspond in comparative relation, such as the root, which is like the mouth, and other things of this type, which will all be known in plants more easily, if they have first become known in animals, where they are more distinct and more evident.

: VIII :

In which the argument of our opponents is done away with,
and in which it is shown in what way the order of
teaching is from things more known to us.

Now that the opinion has been confuted of those who say the ra- 1 tional way of ordering has to be taken not from our knowledge but from the nature of the things to be treated, it is easy to do away with their argument, which was taken from the first book of the *Physics*. For we say that they accept a false [position] when they say Aristotle wanted to commence with a treatment of beginning-

notiora sint, sed quatenus sunt principia, proinde secundùm naturam anteponenda; iam enim diximus Aristotelem non hac ratione
demonstrasse ordiendum esse à principiis, quia sint natura priora
effectis, sed quia aliorum cognitionem assequi non possumus nisi
ex principiorum cognitione, itaque ex nostra cognitione rationem
illius ordinis accepit, non ex natura rerum.

2   Verùm quia adversarii in eo argumento sumpsere prima principia rerum naturalium non esse nobis notiora compositis et effectis,
sed potius multò ignotiora, quo fit ut minimè verum esse videatur
id, quod nos cum Averroe diximus, non solùm viam doctrinae, sed
ordinem quoque esse à notioribus nobis; ideò ut omnem penitus
tollamus difficultatem, declarare debemus quonam modo ordo
doctrinae dicatur esse à notioribus nobis.

3   In primis notanda est distinctio illa vulgata, cognitio nostra
duplex est, una imperfecta, quam confusam vocant, altera perfecta,
quam vocant distinctam, quae adhuc duplex est, aliqua enim est
simpliciter perfecta, aliqua verò non simpliciter, sed solùm in genere perfecta dicitur, id est, pro conditione ac natura illius disciplinae, ut exempli gratia philosophus naturalis debet habere plenam
et simpliciter perfectam cognitionem omnium metallorum, ut in
eis nihil ipsi maneat cognoscendum, sin minùs, imperfectam ipsorum notitiam° habere dicitur. Faber autem aerarius non tenetur
tantam[20] habere aeris cognitionem, quantam habet philosophus
naturalis, sed satis est si tantam habeat, quanta illi ad opus et ad
artem suam exercendam sufficiat. Haec igitur simpliciter perfecta
aeris cognitio non dicetur, sed tamen perfecta erit in illo genere, id
est pro illius artis conditione.

principles, not insofar as they are more known but insofar as they are beginning-principles, and so have to be placed first according to nature; for we said that Aristotle demonstrated that what has to be ordered has to be done so from beginning-principles, not for the reason that they are by nature prior to effects, but because we cannot secure knowledge of other things except from knowledge of beginning-principles. He therefore accepted the reason for this order not from the nature of things but from our knowledge.

Now since our opponents assumed in this argument that the first beginning-principles of natural things are not more known to us than composites and effects, but rather more unknown (from which it happens that what we said with Averroës — that not only the way of teaching but the order too is from things more known to us — would not appear to be true at all), then, so as to completely get rid of every problem, we ought to make clear in what way the order of teaching is said to be from things more known to us.   2

In the first place, this commonplace distinction has to be noted: our knowledge is of two types — one imperfect, which [people] call confused, the other perfect, which they call distinct and which is itself of two types — one is perfect absolutely, the other is not perfect absolutely but is said to be perfect only in a genus, that is, by what is characteristic of and the nature of the discipline. Just as, for the sake of example, the natural philosopher ought to have full and absolutely perfect knowledge of all metals, so that nothing about them remains [unknown] to him; if [he knows] less [than that], it is said he has imperfect knowledge° of them. But an artisan in copper is not held to have the same knowledge of copper that the natural philosopher has. It is enough if he has as much as suffices for him to practice his work and his art. This therefore will not be called absolutely perfect knowledge of copper, but nevertheless it will be perfect in that genus, that is, by what is characteristic of that art.   3

4   Praeterea verò quoniam dictio illa 'notius' comparativa est et refertur ad aliud, quod ignotius dicitur, est etiam sciendum° quòd duobus modis contingit aliquod esse altero notius, aut enim ita illo notius est, ut ad illius cognitionem adipiscendam conferat; aut ad illud cognoscendum nihil confert, ut triangulum est nobis notius primo motore aeterno, nihil tamen ad primi motoris cognitionem cognitio trianguli conducit. Quod quidem membrum est à nobis prorsus dimittendum, quoniam, ut omnes concederent, neque in ordine neque in methodo locum habet haec acceptio notioris.

5   Illud autem notius, quod ad alterius ignotioris cognitionem conducit, in tria membra dividi potest, aliquod enim est notius aliquo tanquàm causa cognitionis illius, quia huius cognitio facit nos assequi illius cognitionem sive distinctam sive confusam, id est ut sit medium, per quod alterum potest syllogismo colligi et fieri notum, ut generatio est nobis notior, quàm prima materia, et est causa cognitionis ipsius materiae tanquàm medium nobis notius. Aliquod autem est notius altero non ut causa faciens cognitionem illius, sed ut necessarium pro illius cognitione consequenda, quare vocari potest causa sine qua non, veluti animal est notius equo, quia necessaria est cognitio animalis pro cognitione equi saltem distincta, confert igitur cognitio animalis ad equum cognoscendum, quia nunquàm equus perfectè cognosceretur ignorato animali, et species ignorato genere, attamen non facit genus cognitionem speciei, ex genere enim non infertur species.

6   Aliquod demum ad alterius notitiam° conferre dicitur non quòd eius cognitio sit causa cognitionis illius, neque etiam causa sine qua non, id est ut sit necessarium ad illius habendam cognitionem

And moreover, since this locution, "more known," is compara- 4
tive and is in reference to something else which is called less
known, so it also has to be understood° that something is more
known than something else in two ways. For either it is more
known than the other in that it contributes to obtaining knowledge of it; or it does not contribute to knowing it, just as triangle
is more known to us than the eternal first mover, yet knowledge of
triangle contributes nothing to knowledge of the first mover. This
[latter] branch [of the division], of course, has to be utterly left
aside by us, since, as everyone has conceded, this [latter] meaning
of "more known" has a place neither in order nor in method.

Now the more known that contributes to knowledge of the 5
other, less known, can be divided into three branches. First, something is more known than something else inasmuch as it is the
cause of knowledge of the other, because knowledge of the one
makes us attain knowledge, either distinct or confused, of the
other. That is, it may be the middle [term], by means of which the
other can be gathered using a syllogism and become known, just as
generation is more known to us than first matter and, as a middle
[term] more known to us, is the cause of knowledge of the matter
itself. Second, something is more known than another not as a
cause producing knowledge of the other but as something necessary for gaining knowledge of it. Therefore it can be called the
cause *sine qua non*, just as animal is more known than horse, because knowledge of animal is necessary for knowledge, at least
distinct [knowledge], of horse. Knowledge of animal therefore
contributes to knowing horse, because horse would not be perfectly known if animal were unknown — if the genus is unknown,
so the species. The genus cannot, however, bring about knowledge
of the species, for the species cannot be inferred from the genus.

Lastly, something is said to contribute to knowledge° of some- 6
thing else not because knowledge of it is the cause of knowledge of
the other, and not because it is a cause *sine qua non* (that is, [not

ipsum cognoscere; sed solum quia est utile, eo enim cognito, facilius consequimur cognitionem alterius. Hac ratione animalia dicuntur nobis notiora, quàm stirpes, licèt ex animalibus nihil possimus inferre de stirpibus, neque necessarium sit animalia cognoscere ad stirpium cognitionem adipiscendam, possunt enim stirpes cognosci, etiam distincta cognitione, animalibus ignoratis, quia sunt duae distinctae species eiusdem generis planta et animal, quarum neutram necessarium est cognoscere pro cognitione alterius. Attamen facilius plantas cognoscimus, si prius animalia cognoverimus, ob eam, quam suprà tetigimus, rationem. Similiter potest cognosci sensus tactus sine cognitione visus, magna tamen cum difficultate. At visu aliisque sensibus cognitis cognitionem tactus facilius adipiscimur.

7   Ex his tota ordinis natura manifesta fit, necnon differentia inter ordinem et methodum propriè sumptam, quae via doctrinae nominari solet. Methodus enim est progressus à notioribus iuxta primam acceptionem notioris, nempè ex quibus inferantur illa ignota, quae quaeruntur. Ordo autem est processus à notioribus iuxta secundam ac tertiam acceptionem, non iuxta primam, nisi ex accidenti, ut infrà declarabimus. In ordine enim illud semper animadvertimus ut ab iis incipiamus, quae vel necessarium vel saltem utile est cognoscere, ut alia, quae cognoscenda manent, melius et facilius cognoscantur. In ordine igitur nullam facimus illationem, nullam argumentationem, sed solam dispositionem rerum tractandarum, ut hanc prius tractemus, illam verò posterius.

8   Unde aliud discrimen oritur inter ordinem et methodum, uterque enim est processus ab hoc ad illud, sed quia methodus est processus illativus; ideò tota tractatio nominatur ab eo, quod

because] it is necessary to know it in order to have knowledge of the other), but only because knowledge of it is useful. For it being known, we more easily gain knowledge of the other. For this reason, animals are said to be more known to us than plants. Even granted that we can infer from animals nothing about plants, neither is it necessary to know animals to obtain knowledge of plants. For plants can be known, even with distinct knowledge, while animals remain unknown, because plant and animal are two distinct species of the same genus and knowing one is not necessary for knowledge of the other. But nevertheless, we know plants more easily if we have first known animals, for the reason we touched on above. Similarly, the sense of touch can be known without knowledge of sight, yet with great difficulty. But if sight and the other senses are known, we obtain knowledge of touch more easily.

From all this the whole nature of order becomes manifest, as 7 does the difference between order and method taken properly, what is normally called the way of teaching. For method is progression from things more known according to the first meaning of "more known," that is, [a progression from the things] from which are inferred the unknown things that are inquired after. Order, on the other hand, is proceeding from things more known according to the second and third meaning, not according to the first, unless by accident, as we will make clear below. For in order we always note something, so that we start from those things that are either necessary or at least useful to know, so that other things, which remain to be known, may be known better and more easily. In order, therefore, we make no inference, no argumentation, but only a disposition of the things to be treated, such that we treat this first and that later.

From this another discriminating difference between order and 8 method arises. Each is a proceeding from this to that. But because method is an inferential procedure, a whole tract is named after that which is inquired after, not after that from which it is

quaeritur, non ab eo, ex quo infertur, hoc enim non propter se, sed illius gratia sumitur, ut in primo libro Physicorum Aristoteles invenit materiam primam methodo resolutiva, nam ex generatione, quae est effectus posterior, collegit esse necessarium dari ipsam primam materiam. Tractatio illa non dicitur esse de generatione et de materia, sed tota de materia, de generatione enim non agitur ibi propter se, sed propter materiam, idcirco Averroem errasse credimus, qui in commentario 57 primi libri Physicorum numeravit eum librum inter libros tractantes de generatione, dicens Aristotelem in eo agere de generatione amplissimè sumpta, quae est unius ex uno, in libro autem de ortu et interitu considerare eam generationem, quae est unius ex pluribus. Attamen generatio non dicitur tractari nisi ubi per se ex suis causis cognoscenda proponitur, et ubi definitur, quare primus liber, in quo Aristoteles de generatione amplissimè sumpta loqui incipiat, est liber de ortu et interitu, ut[21] ex illius libri exordio cognoscere possumus. Et in definitione generationis, quae ibi ab Aristotele traditur, manifestum est Aristotelem de illa loqui, quae est unius ex uno, et de hac multa dicere antequàm de mistione loquatur. Primus autem liber Physicorum non dicitur liber de generatione, sed de principiis solùm, generatio enim ibi sumitur ut leviter et confusè nota, ut ex ea ducamur in principiorum cognitionem.

9   In ordine autem cùm ab hoc ad illud sine illatione progressio fiat, nominatio fit ab utroque, ut si prius agatur de elementis, postea de mistis, non dicetur tota tractatio esse de mistis, sed illa quidem[22] de elementis, haec verò de mistis, ambo enim tractari

inferred. For the latter is taken up not for its own sake, but for the sake of the former, just as in the first book of the *Physics* Aristotle discovered first matter by the resolutive method.[35] For from generation, which is a posterior effect, he gathered that it is necessary that there be first matter itself. The tract is not said to be on generation and matter but wholly about matter, for generation is not dealt with there for its own sake, but for the sake of matter. Therefore we believe Averroës erred when in commentary 57 to the first book of the *Physics*[36] he counted that book among the books treating generation, saying that in it Aristotle deals with generation taken in the widest sense, which is [generation] of one from one; but that in the book on coming to be and passing away, he [i.e., Aristotle] considers generation that is of one from many. But nevertheless, generation is not said to be treated except where what has to be known is set out *per se* from its own causes and where it is defined. And so the first book, in which Aristotle starts to speak of generation taken in the widest sense, is the book on coming to be and passing away, as we can know from the beginning of that book. And in the definition of generation that is conveyed by Aristotle there, it is manifest that Aristotle speaks of that which is of one from one. And he says much about this before speaking about mixture. Moreover, the first book of the *Physics* is not said to be a book about generation, but only about beginning-principles. For generation is taken there as something known lightly and confusedly, so that from it we may be led to knowledge of beginning-principles.

But now in order, since progression occurs from this to that without inference, naming occurs from both, just as, if elements are handled first and then mixed [bodies] afterward, the whole tract is not said to be on mixed [bodies] but the former on elements and the latter on mixed [bodies]. Both are said to be treated

9

seorsum dicuntur et ambo propter se, elementa enim per se cognoscenda sunt, non solùm propter mista; sed anteponitur tractatio de elementis, quia necessaria est ad cognitionem mistorum consequendam.

10　　Evenit autem aliquando ut illa, quae ordinantur, sint eiusmodi, ut illatio fieri possit unius ex alio, propterea quòd reciprocantur, id tamen ex accidenti est respectu ipsius dispositionis, haec enim nullam illationem facit quatenus est dispositio, ut Aristoteles voluit prius agere de natura, postea de motu, et ex natura potest motus inferri, nulla tamen illatio in illa ordinatione consideratur, sed solùm, ut prius de natura agatur, quàm de motu. Nam saepe[23] contingit, ut quod prius tractatur, sit genus, quod verò posterius, sit species, ubi neque consideratur neque fieri potest illatio ulla.

11　　In ordine autem resolutivo ex notione finis videtur illatio fieri eorum, quae ad finem ducunt. Tamen resolutivus id non habet quatenus est ordo, oporteret enim omnem ordinem esse eiusmodi, sed potius quatenus est talis ordo, est enim peculiaris conditio ordinis resolutivi. Sed de hac re loquemur inferius. Satis est in praesentia discrimen inter ordinem et methodum declarasse, quia est essentialis conditio methodi ut huius rei ex illa faciat illationem, ordinem verò, quatenus ordo est, dicimus non esse argumentationem et nullam alicuius rei ex alia re illationem facere secundùm propriam ipsius ordinis naturam.

separately and both for their own sake. For the elements have to be known *per se* and not only for the sake of mixed [bodies], but a tract on elements is placed first, because it is necessary for gaining knowledge of mixed [bodies].

It sometimes happens, however, that those things that are ordered are of the type that inference can be made of [either] one from the other. They can thus be reciprocated. This, however, is by accident with regard to the disposition itself. For this [disposition], insofar as it is a disposition, makes no inference, just as Aristotle wanted to deal first with nature and afterward with motion, and motion can be inferred from nature. So in that ordering, no inference is considered; it is rather only that nature is dealt with prior to motion. For it often happens, where an inference is neither considered nor can occur, that what is treated first is the genus and what is treated afterward is the species.

It appears, however, that in resolutive order, inference of those things that lead to the end arises from a notion of the end. Nevertheless, the resolutive does not have this [feature] insofar as it is an order, for then every order would have to be of this type, but rather insofar as it is a kind of order, for this is a characteristic peculiar to resolutive order. But we will speak on this issue below. At present it is sufficient that we have made clear the discriminating difference between order and method, because the essential characteristic of method is that it makes an inference of this thing from that, and we say that order, insofar as it is order, is not argumentation and, according to the proper nature of order itself, makes no inference of one thing from another thing.

10

11

57

: IX :

In quo et ordinis et methodi utilitas declaratur et clarius
exponitur quomodo ambo sint à notioribus, ad plenam
argumenti contrarii solutionem.

1 Ex his,[24] quae dicta sunt, possumus tam ordinis quàm methodi
utilitatem finemque colligere, cuiusque[25] enim instrumenti natura
in fine et utilitate consistit. Methodi quidem utilitas et finis est
notificare, id est cognitionem facere eius, quod ignoratur, facit igitur methodus ut discamus, quia aperit illud, quod absconditum
erat. Ordo autem hanc vim non habet, nam si res tractandas recto
ordine disponeremus, nec aliqua uteremur methodo vel argumen-
tatione, nihil disceremus et nullam cognitionem adipisceremur.
Sed utilitas ordinis est ut per eum melius et facilius doceamur. Ut
enim materiam primam, et elementa et mista cognoscamus, methodus nobis praestat, non ordo; ab ordine autem habemus ut
melius vel facilius haec omnia discamus. Melius enim ea cognoscimus, si prius agamus de materia prima, mox de elementis,
postea de mistis; quia nisi hunc ordinem servemus, fieri non potest ut perfectam et distinctam horum scientiam habeamus. Similiter quòd sensum visus et sensum tactus cognoscamus, methodo,
non ordini acceptum ferimus. Ut autem facilius horum cognitione
potiamur, ordo praestat. Significavit hoc Aristoteles in principio
quinti libri Metaphysicorum, quando de ordine loquens dixit principium doctrinae non semper esse principium rei, sed unde aliquis
facilius discat, hoc enim ordinis proprium est, quia ordinatè dis-
cendo facilius discimus.

: IX :

In which, toward a full solution to contrary arguments, the
utility both of order and of method is made clear and in what
way both are from things more known is clearly laid out.

From all this that has been said, we can gather the utility and the   1
end of order as well as of method, for the nature of any instrument consists in [its] end and utility. The utility and end of
method, of course, is to make known, that is, to bring about
knowledge, of that which is unknown. Method therefore makes it
that we learn, because it makes apparent that which was hidden.
Order, on the other hand, does not have this power. For if we
dispose the things being treated using the true order, but do not
use some method or argumentation, we would learn nothing and
obtain no knowledge. The utility of order is, instead, that by
means of it we are taught better and more easily.[37] For us to know
first matter, elements, and mixed [bodies], method, not order, is
better for us; but it is by order that we have it that we learn all
these things better or more easily. For we know them better if we
deal first with first matter, next with elements, and afterward with
mixed [bodies], because unless we maintain this order, it is not
possible for us to have perfect and distinct scientific knowledge of
them. Similarly, that we may know the sense of sight and sense of
touch, what is accepted we acquire by method, not by order. But
to get hold of knowledge of these things more easily, order is better. Aristotle indicated this in the beginning of the fifth book of
the *Metaphysics*.[38] When speaking of order, he said that the beginning (*principium*) of teaching is not always the beginning-principle
(*principium*) of the thing but that from which someone may learn
more easily. For this is proper to order, because by learning
[things] in order, we learn more easily.

2   Diversis igitur modis ordo et methodus ad cognitionem conferunt, ideò ambo debuerunt esse progressus à notioribus, alio tamen et alio modo, ut declaravimus.

3   Id verò est hac in re summoperè animadvertendum, quòd cùm
ambo nostram cognitionem tanquàm finem respiciant, non tamen
eandem, methodus enim tam ad confusam quàm ad distinctam
nostram cognitionem dirigitur, ut tunc clarum erit, cùm de methodis locuti erimus, ideò methodus est à notioribus nobis tum
cognitione distincta tum etiam cognitione confusa. Ordo verò
distinctam nostram cognitionem semper respicit, nunquàm confusam. Distinctam intelligo vel simpliciter vel saltem in genere et pro
viribus illius facultatis, quae disponenda est. Ita namque cuiusque
disciplinae partes sunt disponendae, ut optima earum rerum cognitio habeatur, quantum in ea disciplina haberi potest.

4   Ordo igitur semper servatur à notioribus nobis secundùm nostram distinctam et in eo genere perfectam cognitionem. Illa enim,
quae ad aliorum distinctam cognitionem necessarium est cognoscere, prius tractanda sunt, quia iis ignoratis non possunt[26] alia
cognosci, nisi fortasse imperfectè et confusè. Ob hanc rationem
necessarium est prius de genere tractare, quàm de specie, quia distincta et perfecta speciei cognitio tunc habetur, quando eius definitio cognoscitur, in hac autem necesse est accipere genus distinctè
cognitum, quare tractationem de genere tractationi de specie praecedere necesse est.

5   Contingere autem potest ut duo talem inter se respectum habeant, ut neutrum sit necessarium cognoscere ad habendam cognitionem alterius, cùm utrumlibet distinctè cognosci possit etiam
altero ignorato, quod contingit in speciebus, quae aequè sub eodem genere continentur. Tamen alterum, si prius cognoscatur,

Order and method therefore contribute to knowledge in differ- 2
ent ways. Because of this, they both ought to be a progression
from things more known, [but] one in one way, one in another, as
we made clear.

But about this issue the following has to be very diligently 3
noted. Although both [order and method] regard our knowledge
as the end, it is not the same [knowledge in both cases]. For
method is directed as much toward our confused as toward our
distinct knowledge, as will be clear once we have spoken of methods. Because of this, method is from things more known to us
both from distinct knowledge and also from confused knowledge.
Order, on the other hand, always regards our distinct knowledge,
never our confused. I understand "distinct" either absolutely or at
least within a genus, and according to the powers of the branch of
learning[39] that is being disposed. For indeed the parts of any discipline are to be disposed so that optimal knowledge of these things
may be had, to the extent that it can be had in that discipline.

Order therefore is always maintained from things more known 4
to us according to our knowledge, distinct and perfect in the genus. For those things that it is necessary to know for distinct
knowledge of other things have to be treated first, because if they
are unknown, the others cannot be known except perhaps imper-
fectly and confusedly. For this reason, it is necessary to treat the
genus prior to the species, because distinct and perfect knowledge
of the species is then had when the definition of it becomes
known. In this case it is necessary to grasp the genus distinctly
known. And so a treatment on the genus necessarily precedes a
treatment on the species.

Now it can happen that two things have such a relationship 5
with each other that it is not necessary to know neither one to
have knowledge of the other, since either can be known distinctly
while the other is unknown. This happens with species that are
equally contained under the same genus. Nevertheless, one, if it is

alterius cognitionem faciliorem reddat; ut cognitio partium hominis faciliorem reddit cognitionem partium aliorum animalium; et cognitio sensuum perfectiorum faciliorem reddit cognitionem aliorum imperfectiorum, qui difficillimi cognitu sunt, ideò tunc quoque illud anteponere debemus, quod ad aliorum cognitionem confert° tanquàm utile, etsi non confert° tanquàm necessarium.

6   His igitur duabus dictionibus 'melius' et 'facilius' totam ordinis utilitatem exprimimus, ratio enim ordinandi sumitur è nostra cognitione vel melius vel facilius acquirenda; melius quidem, quando aliter[27] disponendo cognitio optima in eo genere haberi non potest; facilius verò, quando potest quidem optima acquiri cognitio etiam illo ordine non servato, at magno cum labore ac difficultate.

7   Duplex igitur est ratio ordinandi omnes disciplinas sumpta ex nostra cognitione tanquàm ex causa finali. Cuiusque enim disciplinae partes, quae rectè sint ordinatae, vel propter perfectae et distinctae scientiae necessitatem vel propter maiorem doctrinae facilitatem ita dispositae sunt. Prior quidem ratio est magis praecipua, et ubi ea locum habet, altera non consideratur. At si illa desit, adsit autem secunda, haec attendi debet. Quòd si neutra locum habeat in quibusdam, in arbitrio authoris positum est à quocumque voluerit exordiri, tunc autem solet ordinem servare nobilitatis vel aliud quippiam eiusmodi. Nullum tamen horum attenditur, quando ratio ordinandi potest ex nostra cognitione desumi altero duorum modorum, quos declaravimus.

8   Fecit autem horum duorum modorum mentionem Aristoteles, nam in prooemio primi libi Physicorum tetigit priorem rationem, quando dixit perfectam rerum naturalium scientiam haberi non posse primis principiis ignoratis, è quorum cognitione pendet perfecta aliorum cognitio, proinde à primis principiis auspicandam

known first, renders knowledge of the other easier, just as knowledge of the parts of man renders knowledge of the parts of other animals easier, and knowledge of the more perfect senses renders easier knowledge of the other, more imperfect ones, which are known with the greatest difficulty. Because of this, then, we also ought to place first that which leads° usefully to knowledge of other things, although it does not lead° necessarily.

With these two locutions, "better" and "more easily," therefore, 6 we express the whole utility of order, for the rational way of ordering is taken from either better or easier acquisition of our knowledge: better, of course, when optimal knowledge in the genus cannot be had by disposing otherwise, and easier, of course, when optimal knowledge can be acquired without maintaining that order, but only with great effort and difficulty.

The rational way of ordering all disciplines therefore is of two 7 types, taken from our knowledge as from a final cause. For the parts of any discipline that are ordered correctly are disposed either on account of the necessity of perfect and distinct scientific knowledge or on account of greater ease in teaching. The first way, of course, is more important, and where it has a place, the other is not considered. But if it is absent and the second is present, the latter ought to be observed. And if in some cases neither has a place, then whatever he should want to begin with is left to the author's[40] choice. He then normally maintains the order of nobility or something else of that type. Nevertheless, none of these is observed when the rational way of ordering can be drawn from our knowledge by either of the two ways that we made clear.

Now Aristotle made mention of these two ways. For he touched 8 on the first way in the proem of the first book of the *Physics*[41] when he said that perfect scientific knowledge of natural things cannot be had if first beginning-principles are unknown. On the knowledge of these, perfect knowledge of the others depends. And so one should commence with first beginning-principles. Who would

esse. Quis autem non videat Aristotelem ibi rationem eius ordinis sumere à perfectae nostrae cognitionis necessitate?

9   Alterius rationis meminit Aristoteles[28] in principio quinti Metaphysicorum, dum dixit in disciplinis non semper principium doctrinae esse principium rei, sed quandoque esse id, unde quispiam facilius discat.

10   Tetigit etiam utramque rationem Averroes in primo capite primi libri Meteorologicorum, ut eo in loco videre possumus.

11   Ex his patet eos errasse, qui putarunt, omnem scientiam et omnem artem posse diversis ordinibus pro scriptoris seu doctoris arbitrio tradi ac disponi, revera enim unus est, non plures, in singula disciplina tradenda ordo melioris ac facilioris doctrinae, qui si pervertatur, certè ea facultas benè ordinata dici non potest, quia sic vel perfecta cognitio acquiri nequit vel difficilius acquiritur. Ideò cui compositivus ordo benè aptatur, illi resolutivus competere non potest, et è converso. Sed hac de re fusius loquemur, cùm de singulis ordinibus sermonem faciemus.

12   Ad argumentum igitur adversariorum clara est responsio, quae omnem difficultatem tollit. Dicebant enim, Aristoteles à principiis exorditur, quae maximè incognita sunt, et composita et effecta sunt iis multò notiora, ergo non est ordinis conditio ut sit à notioribus nobis.

13   Inest in hoc argumento fallacia à secundùm quid ad simpliciter, nam principia esse nobis ignotiora effectis verum est secundùm nostram confusam cognitionem. Simpliciter tamen ignotiora non sunt, quia secundùm ordinem cognitionis distinctae sunt notiora et omnino prius cognoscenda, quàm effectus, quia magis cognoscibilia sunt. Possunt enim distinctè cognosci absque distincta

not see that Aristotle here takes the reason for his order from the necessity of our perfect knowledge?

Aristotle recalls the other way in the beginning of the fifth 9 [book] of the *Metaphysics*[42] where he said that in disciplines the beginning (*principium*) of teaching is not always the beginning-principle (*principium*) of the thing, but sometimes it is that from which anyone may learn more easily.

Averroës too touched on each way in the first chapter of the 10 first book of *On Meteorology*,[43] as we can see in that passage.

From all this it is patent that those who held that every science 11 and every art could be conveyed and disposed with different orders, according to the choice of the writer or teacher, erred. For in each discipline to be conveyed, there is in truth one order, not many, for better and easier teaching. If this is perverted, the branch of learning certainly cannot be called well ordered, because then perfect knowledge either cannot be acquired or is acquired with difficulty. Because of this, resolutive [order] cannot be applied to that to which compositive order is applied well, and vice versa. But we will speak about this issue at greater length when we discourse about each order.

To the argument of our opponents, therefore, there is a clear 12 response that gets rid of all the problems. For they said that Aristotle begins from beginning-principles that are mostly unknown; and composites and effects are much more known than those; therefore, it is not a characteristic of order that it is from things more known to us.

In this argument is the fallacy *a secundum quid ad simpliciter* ("from a certain respect to absolutely").[44] That beginning-principles are less known to us than effects is true—according to our confused knowledge. But absolutely, they are not less known, because according to the order of distinct knowledge they are more known and have to be altogether known prior to effects, because they are more knowable. For they can be known distinctly

effectuum cognitione. At effectus non possunt distinctè cognosci, nisi habita prius distincta cognitione principiorum. Cognitio enim effectus pendet à cognitione causae, sed cognitio causae non pendet à cognitione effectus.

14   Itaque si ratio ordinandi sumeretur à nostra cognitione confusa, utique ratio ipsorum aliquid roboris haberet. Sed cùm à sola distincta sumatur, vana prorsus est et fallax, quemadmodum diximus.

: X :

Quomodo etiam verè dicatur rationem ordinandi sumi semper ab ipsa rerum natura.

1 Id autem non est silentio praetereundum, quòd horum sententia, si sano modo sumatur, quem ipsi certè non respexerunt, vera esse potest. Cùm enim dicimus rerum naturam, duo possumus intelligere, aut ipsam rerum naturam prout extra animam et extra omnem nostram cognitionem sunt, qua ratione non est verum quòd ratio ordinandi ipsas disciplinas sumatur ab illo ipso ordine, quem res habent secundùm naturam. Aut intelligimus rerum naturam ut à nobis cognoscendarum et prout ad nos cognoscentes referuntur, quo quidem modo non negamus ex natura rerum disciplinarum ordinem desumi, talis enim est rerum ut cognoscendarum natura, ut aliae sint nobis notiores, aliae ignotiores; aliae faciliores cognitu, aliae difficiliores; et aliae ad aliarum cognitionem conferant, contrà verò hae ad illas nequaquàm.

without distinct knowledge of the effects. But the effects cannot be known distinctly unless distinct knowledge of the beginning-principles is had first. For knowledge of the effect depends on knowledge of the cause, but knowledge of the cause does not depend on knowledge of the effect.

And so if the rational way of ordering were taken from our 14 confused knowledge, then by all means the reasoning of the others would have some weight. But since it is taken from distinct [knowledge] alone, it is utterly vain and fallacious, as we said.

: X :

In what way it truly is said that the rational way of ordering is always taken from the very nature of things.

Now it should not be passed over in silence that their position can 1 be true, if it is taken in a sound way, a way to which they themselves certainly gave no regard. For when we say the nature of things, we can understand two [things]. Either [we understand it as] the nature itself of things, as they are outside the soul and outside all our knowledge; by this reasoning it is not true that the rational way of ordering the disciplines themselves is taken from the very order that things have according to nature. Or we understand the nature of things as they have to be known by us and as they are referred to us knowing them; [understood] in this way, of course, we do not deny that the order of disciplines is drawn from the nature of things. For such is the nature of things as they have to be known, that some are more known to us, others more unknown; some easier to know, others more difficult; and some contribute to knowledge of others, and on the contrary some do not [contribute to knowledge of] others at all.

2   Hic autem ordo non semper est illemet ordo, quem res secundùm se habent, sed contingere potest hunc illi contrarium esse, ut
alius sit ordo rerum naturalis prout secundùm se considerantur,
alius verò earundem ut optimè et facillimè à nobis cognoscendarum. Ideò cùm talis sit natura nostrorum cognoscentium, ut ab iis
exordiri velimus ac debeamus, quorum cognitio ad aliorum perfectam cognitionem necessaria aut saltem utilis est; et cùm disciplinae propter nos tradantur, ut nos eas cognoscamus, sequitur in iis
omnibus ordinandis posteriorem hunc rerum ordinem semper attendi, non alterum priorem. In hoc igitur sensu verum est quòd
ratio ordinandi omnes disciplinas sumitur ex natura rerum. Verum
est etiam quòd ea semper sumitur à nostra cognitione, haec enim
duo in eundem sensum cadunt, ut considerantibus manifestum
est.

: XI :

In quo definitio ordinis ex iis,
quae dicta sunt, colligitur.

1   Per haec, quae hactenus demonstravimus, conspicua facta est ordinis utilitas et finalis causa. Quoniam igitur ordo instrumentum
nostrae cognitionis est et omnis instrumenti natura in fine et utilitate consistit, ideò si finem hunc in ipsa ordinis definitione expri-
mamus, perfecta erit et omnibus numeris absoluta definitio, quae
alioquin est manca et imperfecta, qualis ea fuit, quam alii attulerunt.

Now the latter order is not always that former order that things 2
have in and of themselves. Indeed it can happen that the latter is
the contrary of the former, so that the one is the natural order of
things as they are considered in and of themselves, and the other
of the same things as they have to be known optimally and most
easily by us. Therefore since such is the nature of our knowing,
that we want and ought to begin with those things, knowledge of
which is necessary or at least useful for the perfect knowledge of
other things, and since disciplines are conveyed for our sake, so
that we may know them, it follows that, in all those things being
ordered, the latter order of things is always observed, not the other
one. In this sense, therefore, it is true that the rational way of ordering all disciplines is taken from the nature of things. But it is
also true that it is always taken from our knowledge. For these two
come down to the same sense, as is manifest to those considering it.

: XI :

*In which the definition of order is gathered
from the things that have been said.*

By means of the things that we have demonstrated up until now, 1
the final cause and utility of order have become plainly apparent.
Since order therefore is an instrument of our knowledge and the
nature of every instrument consists in [its] end and utility, if we
express this end in the very definition of order, the definition will
be perfect and on all counts complete. Otherwise it will be deficient and imperfect, as was the sort that others brought forward.

2   Nos igitur dicimus ordinem doctrinae esse instrumentalem habitum, per quem apti sumus cuiusque disciplinae partes ita disponere, ut quantum fieri possit, optimè ac facillimè illa disciplina discatur.

3   Quòd haec definitio optima sit multa testantur, primùm quidem rei definitae naturam egregiè declarat ac dilucidam reddit, cùm exprimat veram et universalem rationem ordinandi omnes disciplinas, quae quidem est finis et utilitas ipsius ordinis.

4   Omnes etiam difficultates solvit et aliorum errores, errorumque causas aperit. Alii namque non expresserunt quaenam sit vera ordinandi ratio. Quòd si aliquam significasse videntur, in ea falsi° sunt, cùm dixerint rationem ordinandi ex ipsa rerum natura desumendam esse. Quae quidem sententia tantum abest ut dubia solvat, ut potius in plurimas, easque maximas difficultates aliorum animos traxerit. Cùm enim in disciplinis ab illustribus authoribus traditis modò ordinem rerum naturalem, modò ordinem contrarium naturali vel saltem ab eo diversum servatum esse videamus, nullam huius varietatis rationem adducere possumus, dum illam sententiam sequimur et magnis atque insolubilibus difficultatibus urgemur. At nostra opinione recepta omnes difficultates solvuntur, semper enim ostendere possumus ordinem servari nostrae melioris vel facilioris cognitionis, ad quam dum confugimus, causam omnis dispositionis facilè reddimus. Sive enim ordo naturae sive huic contrarius sive alius quilibet servetur, is semper est ordo nostrae melioris cognitionis.

5   Denique haec definitio causam affert omnium eorum, quae ordinem veluti accidentia quaedam consecuntur. Solent omnes dicere, ordo docet, quemadmodum enim confusio sive inordinatio obscuritatem et ignorantiam, ita ordo perspicuitatem et doctrinam parit. Huius autem dicti haec una ratio afferri potest, quia illa

We say, therefore, that order of teaching is an instrumental   2
habit by means of which we are able so to dispose the parts of any
discipline that the discipline may be learned as optimally and easily as can be.

Many things attest that this definition is optimal. First, of   3
course, it makes very clear and renders lucid the nature of the
thing defined, since it expresses the true and universal way of ordering all disciplines. This, of course, is the end and utility of order itself.

It also does away with all the problems and errors of others and   4
makes apparent the causes of [their] errors. For others did not
expressly say what the true and rational way of ordering is. And if
they appear to have indicated one, they were wrong° in it, since
they said that way of ordering has to be drawn from the very nature of things. This position, of course, is so far from doing away
with doubts that instead it has driven the soul of others into many
problems, all large. For when, in disciplines conveyed by illustrious
authors, we see now the natural order of things maintained and
now an order contrary to or at least different from the natural, we
can adduce no reason for this disparity. And when we follow that
position, we are also burdened by great and unsolvable problems.
But once our opinion is accepted, all problems are done away
with. For we can always show that the order for our better or
easier knowledge is being maintained. When we take refuge in this
[opinion], we easily render the cause of every disposition. Whether
the order of nature or its contrary or whatever other [order] is
maintained, it is always the order of our better knowledge.

And lastly, this definition brings forward the cause of all things   5
that ensue from order as accidents of a sort. Everyone normally
says that order teaches: that just as confusion or disorder results in
obscurity and ignorance, so order brings forth perspicuity and
teaching. And for this saying, one reason can be brought forward:

prius tractantur, quae ad aliorum cognitionem conferunt. Dum igitur rationem ordinandi ex nostra cognitione desumimus, causam facilè assignamus cur ordo doceat, eaque negata nullam aliam adducere possumus, cui[29] animus acquiescat.

6   Praeterea undenam collegit multis in locis Aristoteles ordinem doctrinae postulare ut ab universalibus ad particularia procedatur? Certè non aliunde, quàm ex tradita à nobis ordinis definitione, nempè ex nostra meliore cognitione. Quoniam enim cognitio universalium est necessaria ad perfectam cognitionem particularium et cognitio generis ad cognitionem specierum, ideò semper iussit de genere prius agendum esse, quàm de speciebus, et tractationem universalium anteponendam esse tractationi particularium. Certè Aristoteles in prooemio primi libri Physicorum non alia ratione probavit esse ab universalibus ad particularia progrediendum, quàm quia universale, quod de multis praedicatur, est nobis notius particularibus et ita notius, ut necessaria sit ipsius cognitio ad particularium cognitionem perfectam adipiscendam.

7   Averroes quoque in praefatione sua libros illos, necnon in sua praefatione in Posteriores Analyticos tres rationes adducit, cur universalia sint anteponenda particularibus in omni doctrina, quae omnes ex nostra meliore ac faciliore cognitione desumuntur, ut in memoratis locis legere possumus, non est enim operae pretium ea omnia, quae ibi dicuntur ab Averroe, in praesentia recensere, cùm locos ostendisse° satis sit.

it is because those things are treated first that contribute to knowledge of other things. Therefore as long as we draw the way of ordering from our knowledge, we may easily assign the cause why order teaches. And if this is denied, we can adduce no other [position] in which our soul acquiesces.

Moreover, from what did Aristotle in many passages gather that   6
the order of teaching demands that one proceeds from universals to particulars? Certainly not from anything other than the definition of order conveyed by us, that is, from our better knowledge. For since knowledge of universals is necessary for perfect knowledge of particulars and knowledge of the genus for knowledge of the species, he always demanded that the genus be dealt with prior to the species, and a treatment of universals be placed before a treatment of particulars. Certainly, in the proem of the first book of the *Physics*,[45] Aristotle proved that progression has to be from universals to particulars for no other reason than that the universal, which is predicated of many, is more known to us than particulars and so much more known that knowledge of it is necessary for obtaining perfect knowledge of particulars.

Averroës too, in his preface to those books[46] as well as in his   7
preface to the *Posterior Analytics*,[47] adduced three reasons why universals have to be placed before particulars in all teaching, all of which [reasons] are drawn from our better and easier knowledge, as we can read in the passages referred to. It is not worth the work to recall at present everything that was said there by Averroës; it is enough to have pointed out° the passages.

: XII :

Quòd procedere ab universalibus ad particularia
non sit proprium solius ordinis compositivi.

1   Per haec manifestus fit error illorum, qui putant processum ab
universis ad singula proprium esse ordinis compositivi. Nam si
ratio huius processus est sola nostra melior cognitio, haec autem
finis est et communis conditio omnium ordinum, igitur processus
hic naturam ordinis quatenus ordo est consequitur, proinde omni
ordini competit, non soli compositivo.

2   Patet autem Aristotelem non modò in scientia naturali, quae
ordine compositivo tradita est, ab universis ad singula processisse;
verùm etiam in libris moralibus et in Analyticis, in quibus servavit
ordinem resolutivum. In moralibus enim prius de virtute in universum locutus est, postea de virtutibus singulis. In Analyticis autem Posterioribus prius à[30] notione finis invenit medium demonstrationis esse causam, propter quam res est, nomine causae
generaliter accepto in toto primo libro et in principio etiam secundi. Postea verò in eodem secundo distinxit nomen causae in
quatuor genera causarum, ut doceret quomodo eorum quodlibet
in demonstratione potissima medium esse possit.

3   Medicus quoque proponens sibi unumquemque morbum
curandum, prius universales remediorum conditiones investigat,
deinde ad particulares descendit. Ut si ei proponatur tradenda
curatio morbi calidi et sicci, prius decernet utendum esse remediis
frigidis et humidis, deinde ad species frigidorum et humidorum
descendens considerabit quae conveniant et quae non conveniant.

: XII :

That to proceed from universals to particulars is not proper to compositive order alone.

By means of all this, the error of those who hold that proceeding 1
from universal wholes to each [of the parts] is proper to compositive order becomes manifest. For if the reason for this procedure is
only our better knowledge, and this is the end and a common
characteristic of all orders, it ensues, therefore, that this procedure
is the nature of order insofar as it is order, and accordingly it appertains to every order, not just to compositive.

And now it is patent that Aristotle proceeded from universal 2
wholes to each [of the parts] not only in natural science, which is
conveyed using compositive order, but also in the books on ethics
and in the *Analytics*, in which he maintained resolutive order. For
in [the books] on ethics, he spoke first about virtue as a universal
whole and afterward of each virtue. And in the *Posterior* [*Analytics*],
in the whole first book and also in the beginning of the second,
first accepting the name of cause in the general sense, he discovered from a notion of the end that the cause by which a thing is, is
the middle [term] of a demonstration. And then afterward in the
second [book],[48] he divided the name of cause into four kinds of
causes so that he could teach in what way any of them can be a
middle [term] in a demonstration *potissima*.

A physician too, setting out to cure some one disease, first in- 3
vestigates the universal characteristics of remedies and then de-
scends to particulars. Just as, if he is set [the task of] conveying
the cure for a warm and dry disease, he will first decide that cold
and moist remedies have to be used; then, descending to the species of things cold and moist, he will consider which are appropriate and which are not appropriate.

4   Verùm medici hanc nostram appellationem non probant, quod
enim nos universale vocamus, ipsi remedium vocandum esse dicunt, ut frigidum. Quod verò particulare nominamus, ipsi materiam ipsius remedii appellandam asserunt, ut lactucam.

5   Attamen ut nos eorum appellationem non reprehendimus, ipsi
quoque nostram reprehendere non deberent, species enim remedii
frigidi lactuca est, de qua frigidum in casu recto praedicatur; quae
etiam non incongruè materia illius remedii dicitur, quandoquidem
universale, ut Aristoteles ait in primo libro de Coelo, significat
formam, particulare verò formam in materia. Non debent igitur
negare processum illum à frigido simpliciter ad lactucam esse ab
universali ad particulare, quam communem omnis ordinis conditionem esse diximus.

6   Sed res haec magis declarabitur cùm de singulis ordinibus sermonem faciemus. Haec voluimus breviter hoc in loco annotare, ut
pateret communem esse omnis ordinis proprietatem ab universalibus ad particularia progredi, non solius compositivi, ut aliqui arbitrantur.

: XIII :

In quo dubium quoddam proponitur contra ea,
quae proximè dicta sunt.

1   Caeterum adversus ea, quae modò diximus, dubitare non absque
ratione quispiam posset, videtur enim non esse ordinis proprium
ut ab universalibus ad particularia progrediatur, siquidem apud
probatos authores comperimus quandoque speciem generi et particulare universali in tractatione fuisse antepositum. Euclides quatuor elementorum libros quinto anteposuit, in illis tamen de lineis
tantùm ac superficiebus agit, in quinto autem de magnitudine

It is true that physicians do not approve our appellation. For 4
what we call the universal, such as cold, they say has to be called
the remedy. And what we call the particular, such as lettuce, they
assert has to be called the matter of the remedy.

Now we do not censure their appellation, and they ought not 5
to censure ours either. For the species of the cold remedy is lettuce, of which cold is predicated *in casu recto*[49] and which is not
inaptly said to be the matter of the remedy, since the universal, as
Aristotle says in the first book of *On the Heavens*,[50] signifies form,
and the particular [signifies] form in matter. They ought not to
deny, therefore, that this proceeding from cold absolutely to lettuce
is from universal to particular, which we said is a common characteristic of every order.

But this issue will be made clearer when we discourse about 6
each order. In this passage we wanted to note these things briefly,
so it may be clear that to progress from universals to particulars is
a common property of every order, not of only the compositive, as
others think.

: XIII :

In which is proposed a doubt against the
things that have just been said.

But now someone could, not without reason, raise a doubt against 1
the things that we just said. For it appears not to be proper to order that it progress from universals to particulars, since in proven
authors we sometimes find in a treatment that species was placed
before genus and particular before universal. Euclid placed four
books of elements before the fifth, and in them deals only with
lines and surfaces. In the fifth, however, [he deals] with magnitude

amplissimè sumpta, quae et lineas, et superficies et solida quoque complectitur. Unde colligitur, Geometriam et Stereometriam non duas esse scientias omnino distinctas, sed duas eiusdem scientiae partes, in qua magnitudo latissimè sumpta subiectum constituatur. Quintus igitur ille Elementorum liber cùm utramque partem sua communitate comprehendat et in ipso supremo genere subiecto versetur, primus omnium esse debuit, si necessaria omnis ordinis conditio est ut ab universalibus ad particularia progrediatur.

2   Aristoteles quoque in tractatione de rebus viventibus non videtur universalia particularibus anteposuisse. Prius enim de corpore vivente debuit generaliter et communiter loqui, deinde de eius speciebus, nempè animalibus et stirpibus. Tamen primo loco de animalibus egit, omnium enim de viventibus librorum primi sunt libri de Historia animalium, postea libri de Partibus animalium, qui certè animalium proprii sunt. Post eos egit de anima in universum in libris de Anima, quae quidem tractatio omnia viventia complectitur. Quare tractatio particularis fuit anteposita universali.

3   In iis quoque libris, qui Parvi naturales vocantur, idem inspicere possumus. Aliqui enim, qui ad sola animalia pertinent, aliquibus sunt antepositi animantia omnia complectentibus. Prius enim egit Aristoteles de sensu et sensilibus, de memoria et reminiscentia, de motu animalium, quàm de vita et morte, et de longitudine ac brevitate vitae, de iuventute et senectute, quae communia accidentia sunt omnium animatorum. Aut igitur Aristoteles et Euclides non rectè fecerunt, aut non est necessarium in disponendis cuiusque disciplinae partibus procedere ab universalibus ad particularia.

taken in the widest sense, which encompasses lines, surfaces, and also solids. From this it is gathered that geometry and stereometry[51] are not two altogether distinct sciences, but two parts of the same science, in which magnitude taken in the broadest sense constitutes the subject. The fifth book of the *Elements* therefore, since it comprehends both parts by their commonality and is concerned with the highest genus as [its] subject, ought to be the first of all, if a necessary characteristic of every order is that it progresses from universals to particulars.

Aristotle too, in [his] treatment of living things, appears not to 2 have placed universals before particulars. For he ought to have spoken first on living body in the general and common sense, and then on its species, namely animals and plants. But he dealt in the first place with animals. For the first of all the books on living things are the books *On the History of Animals*; afterward are the books in *On the Parts of Animals*, and these are certainly proper to animals. After these[52] he dealt with soul as a universal whole in the books *On the Soul*, a treatment that, of course, covers all living things. Therefore the treatment of particulars was placed before the treatment of the universal.

Also in the books that are called the *Parva Naturalia*, we can 3 observe the same thing. For those that pertain to animals alone are placed before the others covering all animate things. For Aristotle dealt first with sense and sensibles, memory and remembrances, and motion of animals, then with life and death, longevity and shortness of life, and youth and old age; these [latter] are common accidents of all animate things. Therefore either Aristotle and Euclid did what they did incorrectly, or in disposing the parts of any discipline it is not necessary to proceed from universals to particulars.

: XIV :

In quo dubii solutio exponitur.

1 Difficultas haec ad plenum solvi non potest, nisi et de rebus mathematicis et de iis, quae ad animantia pertinent, multa dicamus, quae fortasse limitum tractationis logicae transgressio alicui esse videbitur. Attamen animadvertere quisque debet nos de rebus logicis ita tractationem instituisse, ut non modò logicam à rebus abiunctam, quam docentem vocant, sed applicatam quoque rebus ac disciplinis nobis considerandam proposuerimus. Eius enim, qui logicas regulas plenè cognosciturus sit, non parum interest tum ea, quae de rebus logicis docuit Aristoteles, considerare; tum etiam quomodo et Aristoteles et alii probati authores in disciplinarum traditione praeceptis logicis usi fuerint, diligenter observare. Res igitur naturales vel mathematicas vel medicas declarare consilium nostrum non est. Sed tamen ipsam dispositionem ac methodum, quae in his disciplinis servata est, cum quadam moderatione considerandam à nobis esse censemus, ubi eam ad rerum logicarum intelligentiam prodesse aliis posse iudicaverimus.

2 Dicimus igitur universale particularibus in doctrina esse anteponendum non quia sit universale, sed melioris doctrinae gratia, quoniam cognitio particularium ex universalium notitiaᵒ pendet. Igitur si quandoque contingat ut propter cognitionem melius sit anteponere particulare universali, absque ullo dubio universale ob eandem rationem postponendum est. Ideò praeceptum illud Aristotelis, quòd tractatio universalium debeat praecedere tractationi de particularibus, intelligendum est ut verum in plurimis, aliquando enim contingit (quamvis rarò) ut facilioris doctrinae ratio cogat ordinem hunc aliquantulum perturbare.

: XIV :

In which the solution to the doubt is laid out.

This problem cannot fully be done away with unless we say many   1
things about both mathematical issues and those that pertain to
animate things. This will perhaps appear to some to be a trans-
gression of the bounds of a treatment of logic. But nevertheless
everyone ought to note that we put this treatment on logical issues
together so that we might set out for consideration by us not only
logic separated from things (which they call teaching) but also
[logic] applied to things and to disciplines. For it is of no small
interest to anyone who would fully know the rules of logic both to
consider that which Aristotle taught on logical issues and also to
observe carefully in what way both Aristotle and other proven authors used logical precepts in the conveying of disciplines. It is,
therefore, not our intent to clarify natural or mathematical or
medical issues. But instead we deem that the disposition itself and
the method that is maintained in these disciplines have to be—
with some moderation—considered by us when we have judged that
it could be profitable to others in understanding issues of logic.

We say therefore that in teaching, the universal has to be placed   2
before particulars, not because it is universal, but for the sake of
better teaching, since knowledge of particulars depends on knowl-
edge° of universals. If therefore it sometimes happens that for the
sake of knowledge it is better to place the particular before the
universal, without any doubt the universal has to be placed after
for the same reason. And so, that precept of Aristotle's—that a
treatment of universals ought to precede a treatment on particulars—has to be understood as true [only] in the majority of cases,
for it sometimes happens (even though rarely) that the rational
way of easier teaching forces [us] to disturb this order a little.

3   Quomodo autem et quam ob causam id eveniat, considerandum est.

4   In omni re proposita duo sunt, quae cognoscere volumus,
primùm quidem essentia sive natura ipsius rei; deinde propria eius
accidentia. Ideò quando decernimus agendum esse prius de genere,
postea verò de specie, ipsius quidem[31] generis naturam et accidentia primo loco declaramus, deinde naturam et accidentia speciei.
Ipsa enim natura speciei non potest cognosci, dum natura generis
ignoratur. Declaraturus igitur naturam hominis debet prius animalis naturam declarasse. Hinc colligimus, quòd si fortè contin-
geret, ut tum generis tum speciei natura, vel saltem sola natura
generis esset res per se nota et nulla egeret declaratione, tunc necessarium non esset genus speciei ordine doctrinae anteponere,
nisi cognitio proprietatum id postularet.

5   Accidentium autem propriorum tum generis tum speciei ea est
differentia, quòd aliquando proprietas speciei est species proprieta-
tis generis, aliquando non est eius species, ut motus est proprietas
corporis naturalis, motus autem à medio est proprietas corporis
levis, et quemadmodum corpus leve est species corporis naturalis,
ita motus à medio est species quaedam motus. Calor autem est
eiusdem corporis levis proprietas, quae non est species motus. Ita
aptitudo ad ridendum in homine non est species sensus, qui animalis proprius est. Huius autem differentiae ea est ratio, quòd
propria speciei natura tum proprias edit operationes et accidentia
propria distincta ab accidentibus propriis generis; tum etiam ipsas
generis proprietates coarctat et sibi aequales reddit, ut motus animalis alio modo in ave coarctatur, alio in pisce, et alio in terrestri
animali; sic etiam alio modo in homine et alio in brutis. Propterea

In what way and by what cause this happens has to be considered. 3

In everything set out there are two things that we want to 4 know: first, of course, the essence or nature of the thing itself, and then accidents proper to it. And so when we decide to deal first with the genus and then afterward with the species, we, in the first place, clarify the nature and accidents of the genus itself, of course, and then the nature and accidents of the species. For the very nature of the species cannot be known when the nature of the genus is unknown. Anyone going to make clear the nature of man, therefore, ought first to have made clear the nature of animal. And thus we gather that, if it happened by chance that the nature of both the genus and the species, or at least just the nature of the genus, were known *per se* and needed no clarification, then it would not be necessary to place the genus before the species in the order of teaching, unless knowledge of the properties demanded it.

Now it is a differentia of the proper accidents both of genus 5 and of species that sometimes a property of the species is a species of the property of the genus, and sometimes it is not a species of it. Just as motion is a property of natural body, motion from the middle,[53] moreover, is a property of light body; and just as light body is a species of natural body, so motion from the middle is a species of motion. Heat, however, is a property of the same light body, and this [property] is not a species of motion. And then in man the ability to laugh is not a species of sense, which is proper to animal. The reason for this difference is that the nature proper to the species both brings about proper activities and proper accidents distinct from the proper accidents of the genus and also narrows down the properties themselves of the genus and explains what is equal in total extent to them. Just as the motion of animal is narrowed down in one way in bird, another in fish, and another in terrestrial animal; and yet another way in man and another in

proprius hominis motus est species illius, qui animalis proprietas
est.

6   Quando igitur accidentia generis conferimus cum accidentibus
speciei, quae illorum species sunt, certum est accidentia generis
prius esse cognoscenda, quàm accidentia speciei, motum enim
elementorum et motum animalium benè cognoscere non possumus, nisi prius motum universè acceptum cognoverimus ut proprietatem corporis naturalis. Sic etiam de generatione simpliciter
sumpta prius agendum est, quàm de generatione animalium.

7   Quòd si accidentia speciei non sint species accidentium generis,
nulla doctrinae necessitas cogit ut antè de accidentibus generis,
quàm de speciei accidentibus disseramus. Potest quidem substantia generis, si aliqua declaratione indigeat, postulare ut prius de
genere, quàm de specie tractatio fiat, sed si substantia rei non
considerata,[32] solas affectiones spectemus, nulla certè necessitate
compellimur ad accidentia generis prius tractanda, quàm accidentia speciei; nisi forte facilitate doctrinae ducamur, solent enim accidentia generis faciliora cognitu esse, quàm accidentia speciei,
quando utraque sensilia sunt; tunc enim prius et facilius accidentia
generis, quàm accidentia speciei sentiuntur. In talibus igitur
quando distinctam cognitionem quaerimus, à facilioribus, proinde
à communioribus exordiri solemus. Verùm si neutra sensilia sint,
non modò nulla doctrinae necessitas, sed ne ipsa quidem facilitas
postulat ut accidentia generis prius declaremus, quàm accidentia
speciei. Quod quidem in rebus mathematicis, ut mox declarabimus, potissimùm contingit.

8   Ex his omnibus, quae dicta sunt, hoc elicimus, in quo tota pro-
positae dubitationis solutio consistit, quòd si eveniat ut natura
generis sit per se nota et nulla declaratione indigeat et accidentia
speciei non sint species accidentium generis, non est necessarium

beasts. Thereby, the motion proper to man is a species of that which is a property of animal.

When therefore we compare accidents of the genus with accidents of the species that are their species, certainly accidents of the genus have to be known prior to accidents of the species. For we cannot know the motion of the elements and the motion of animals well, unless we have first known motion accepted universally, as a property of natural body. Thus also, generation taken absolutely should be dealt with prior to the generation of animals. 6

Now if the accidents of the species are not species of the accidents of the genus, no necessity in teaching forces us to discuss accidents of the genus before the accidents of the species. Of course, the substance of the genus can, if it needs some clarification, demand that a treatment on the genus be made prior to that of the species. But if, the substance of the thing not considered, we look only at the affections, we are certainly not compelled by any necessity to treat the accidents of the genus prior to the accidents of the species, unless perhaps we are led [to do so] by ease in teaching. For normally, when both are sensible, accidents of the genus are known more easily than accidents of the species, for then the accidents of the genus are sensed earlier and more easily than the accidents of the species. In such cases, therefore, when we inquire after distinct knowledge, we normally begin with easier things and accordingly with things more common. But if neither is sensible, not only no necessity in teaching but not even ease itself [in teaching] demands that we clarify the accidents of the genus prior to the accidents of the species. This happens chiefly, of course, in mathematical issues, as we will soon make clear. 7

From all the things that have been said, we find the following, in which consists the whole solution to the proposed doubt. Now if it happens that the nature of the genus is known *per se* and is in need of no clarification, and the accidents of the species are not species of the accidents of the genus, then it is not necessary for 8

pro nostra meliore cognitione ut prius de genere, quàm de specie
agatur. Nam de ipsa generis substantia agendum non est, cùm sit
per se nota, eius autem accidentia non conferunt ad cognoscenda
accidentia speciei. Ordo quidem naturalis ipsarum rerum require-
ret ut prius de accidentibus generis ageretur; sed si ordo facilioris
doctrinae postularet ut speciei accidentia praemitterentur, hic certè
servandus esset relicto ordine naturali.

9   Evenit hoc in scientiis mathematicis, Geometria enim prout
Stereometriam tanquàm partem continet, subiectum habet magnitudinem latissimè acceptam, sed mente ab omni sensili materia
abstractam, cuius quidem ita acceptae substantia et essentia per
solam nominis declarationem tota exprimitur et ab omnibus intel-
ligitur, est enim simplex quoddam accidens, ad cuius essentiam
significandam sola vocabuli declaratio sufficit, proinde nominalis
dicta definitio ab essentiali definitione non differt; neque perfec-
tior ibi datur definitio, quàm nominalis, ideò rectè dicebat Averroes in commentario secundo primi libri Posteriorum declarati-
ones nominum rerum mathematicarum effici perfectas in eo genere
definitiones.

10   Geometra igitur tales definitiones tanquàm principia profert,
quia simulatque audiuntur, intelliguntur et sunt per se notae.
Quandoque etiam eas non profert, quia sine ulla expressione sunt
omnibus notissimae, qualis est definitio magnitudinis latè sumptae, omnes enim norunt magnitudinem esse id, quod aliquà sit
dimensum. Propterea acturus de magnitudine Euclides in quinto
Elementorum libro eius definitionem exprimere neglexit, sed nomina tantùm quorumdam eius accidentium demonstrandorum
declaravit. Totus igitur ille quintus liber est de accidentibus magnitudinis amplè acceptae, de ipsius autem substantia et natura nihil
dicitur.

our better knowledge that the genus be dealt with prior to the species. For the substance itself of the genus need not be dealt with since it is known *per se*; moreover its accidents do not contribute to knowing the accidents of the species. Of course, the natural order of the things themselves may require that the accidents of the genus be dealt with first, but if the order for easier teaching demands that the accidents of the species be put forward first, this certainly has to be maintained, with the natural order remaining to the side.

This happens in mathematical sciences. For, in that geometry   9
contains stereometry as a part, so it has [as its] subject magnitude accepted in the broadest sense, but abstracted by the mind from all sensible matter. So accepted, of course, its substance and essence are, by means of a clarification alone of the name, wholly expressed and understood by everyone. For it is some sort of simple accident, for which clarification alone of the term is sufficient for signifying its essence. Accordingly, the so-called nominal definition does not differ from the essential definition, nor is there a definition more perfect than the nominal. Because of this, Averroës correctly said in the second commentary to the first book of the *Posterior* [*Analytics*]⁵⁴ that clarifications of the names of mathematical things become perfect definitions in the genus.

A geometer therefore advances such definitions as beginning-   10
principles, because as soon as they are heard, they are understood and are known *per se*. Sometimes he does not even advance them, because [even] without any express statement they are very well known to everyone. The definition of magnitude taken in the broad sense is of this sort. For everyone knows that magnitude is that which is in some way measured. Therefore Euclid, going on to deal with magnitude in the fifth book of the *Elements,* neglected to expressly state a definition of it. He instead only clarified the names of some of its accidents being demonstrated. The whole fifth book therefore is about the accidents of magnitude accepted in a wide sense; nothing is said about its substance and nature.

11 Species autem magnitudinis similiter sunt per se notae, et earum definitiones per se intelliguntur, exprimendae tamen fuere, ut earum discrimen nosceretur,° quod in dimensionum diversitate consistit.

12 De his igitur tractaturus Euclides potuit à speciebus magnitudinis tractationem auspicari, quoniam ad earum essentiam intelligendam nulla generis declaratio requirebatur, cùm sit per se notum et nullibi ab Euclide declaretur. Accidentia verò specierum magnitudinis non sunt species accidentium generis, quare demonstrari et cognosci possunt etiam ignoratis accidentibus generis, videmus enim nullam esse demonstrationem in quatuor anterioribus libris Euclidis, quae supponat aliquam earum demonstrationum, quae in quinto libro fiunt; sicuti neque ulla demonstratio quinti libri pendet ex aliqua earum, quae in illis quatuor libris factae fuerant. Unde patet nullam doctrinae necessitatem coegisse Euclidem aut à primo, aut à quinto libro exordiri. Poterat enim ab utrolibet sumere tractationis initium. Sed tamen propter maiorem facilitatem à primo libro incipere voluit, non à quinto, non quòd primi libri cognitio faciliorem reddat cognitionem quinti, sed quia demonstrationes primi libri sunt simpliciter intellectu faciliores. Quae verò in quinto fiunt, difficillimae. Certum est enim et accidentia magnitudinis universim³³ sumptae et accidentia linearum ac superficierum, quae in iis libris demonstrantur, omnia insensilia esse et nobis penitus incognita non modò cur sint, sed etiam quòd sint. Ideò tantò difficiliora intellectu sunt, quantò universaliora et abstractiora à suppositis, quia non ita possunt illae demonstrationes per materialia exempla ob oculos poni, siquidem exemplo linearum proposito non statim apparet ita esse in superficiebus ac solidis, nisi aliqua mentis adhibita consideratione.

Now although the species of magnitude are similarly known *per*   11
se, and their definitions are understood *per se,* they nevertheless
had to be expressly stated so that the discriminating difference
between them, which consists in the difference of the dimensions,
could be known.°

Euclid therefore in going on to treat these things, could have   12
commenced the treatment with the species of magnitude, since no
clarification of the genus was required for understanding their essence, it being known *per se* and thus nowhere clarified by Euclid.
Indeed the accidents of the species of magnitude are not species of
the accidents of the genus, and so they can be demonstrated and
known while accidents of the genus are unknown. And so we see,
there is no demonstration in Euclid's earlier four books that pre-
supposes any of the demonstrations that are made in the fifth
book, just as no demonstration in the fifth book depends on any
of the ones that had been made in those four books. From this it
is patent that no necessity in teaching forced Euclid to begin either
with the first or with the fifth book. He could have taken the start
of the treatment from either. But nonetheless, on account of
greater ease, he wanted to start from the first book, not from the
fifth, not because knowledge of the first book renders knowledge
of the fifth easier, but because the demonstrations of the first book
are easier to understand absolutely. Those made in the fifth are
very difficult. For it is certain that both the accidents of magnitude
taken universally and the accidents of lines and surfaces that are
demonstrated in those books are all insensible and completely unknown to us: not only why they are but also that they are. They
are as much more difficult to understand as they are more universal and abstracted from [their] subjects, because those demonstrations cannot be placed before the eyes by means of material examples. Even when an example of lines is set out, it is not at once
apparent that surfaces and solids are the same, unless by some
applied consideration by the mind.[55]

13   Quoniam igitur scientiae mathematicae illa tempestate pueris ante solidiores alias disciplinas proponebantur tanquàm faciliores, ut contemplationi assuescerent et in illis demonstrationibus exercerentur, ideò voluit Euclides à faciliore Geometriae parte auspicari; et usque adeò facilitatem doctrinae sectatus est, ut Geometricorum librorum seriem interpositis Arithmeticis libris interrumpere voluerit, nam post sex Elementorum libros Geometricos septimum, octavum et nonum Arithmeticos posuit, in quibus de numeris agit, postea in decimo libro ad magnitudines revertitur et totam Geometriam aliis pluribus libris absolvit, nam Geometriae nomine Stereometriam quoque comprehendimus, de qua postremi Elementorum libri scribuntur.

14   Ex libris igitur Euclidis magna sententiae nostrae confirmatio desumitur, siquidem per eos ostendimus totam ordinandi rationem à nostra meliore seu faciliore cognitione accipiendam esse, semper enim attenditur doctrinae facilitas, dummodò perfectae in eo genere cognitionis necessitas in contrarium non urgeat. Hac autem urgente servatur ordo perfectioris doctrinae, etiam si ordini rerum naturali contrarius sit.

15   Ad alterum dubium de libris Aristotelis de animalibus dicimus non omninò perversum esse in iis libris ordinem ab universalibus ad particularia, sed levem quandam mutationem[34] suscepisse propter eas causas, quas modò declarabimus. Certum est enim tres esse debuisse distinctas tractationes, primam quidem de corpore vivente latè sumpto, secundam de animalibus, tertiam verò de plantis. Eorum enim, quae viventibus competunt, alia animalium propria sunt, alia propria stirpium, alia utrisque communia, quae quidem aliis omnibus erant anteponenda. Tamen Aristoteles propter meliorem ac faciliorem doctrinam non tres tractationes fecit, sed duas, unam[35] de animalibus, alteram de plantis, sed cum

And so, since in that time mathematical sciences were given to   13
children as easier disciplines before [they were given] other, denser
ones, so that they could become accustomed to contemplation and
practiced in such demonstrations, Euclid wanted to commence
with the easier part of geometry, and he followed ease of teaching
so much that he wanted to interrupt the sequence of books on
geometry with an interposition of books on arithmetic. For after
six books of the *Elements* on geometry, he placed the seventh,
eighth, and ninth on arithmetic, in which he deals with numbers.
Afterward, in the tenth book, he returns to magnitude and completed the whole of geometry in many other books. For under the
name of geometry we also comprehend stereometry, on which the
final books of the *Elements* are written.

Large confirmation for our position therefore can be drawn   14
from Euclid's books, since by means of them we showed that the
whole rational way of ordering has to be accepted for our better or
easier knowledge. For ease of teaching is always observed as long
as the necessity of perfect knowledge in the genus does not urge
[us] in the contrary direction. But even when this does urge [us],
the order of more perfect teaching is maintained, even if it is contrary to the natural order of things.

To the other doubt, regarding Aristotle's books on animals, we   15
say that the order from universals to particulars is not entirely
perverted in these books;[56] instead [he] undertook a sort of light
change, on account of causes that we will now make clear. For it is
certain there ought to have been three distinct treatments, the
first, of course, about living body, taken in a broad sense, the second about animals, and the third about plants. For of those
[qualities] that appertain to living things, some are proper to animals, some proper to plants, and others are common to both,
which [last], of course, should have been placed before all the others. But Aristotle, for the sake of better and easier teaching, made
not three but two treatments, one on animals, the other on plants.

tractatione de animalibus commiscuit illam, quae est de communibus. In animalibus namque tum illa, quae animalium propria sunt, tum illa, quae sunt eis communia cum plantis, considerare constituit. Non quòd haec communia attribuat animalibus, ut animalia sunt, perinde enim esset ac demonstrare triangulum aequilaterum tres habere angulos aequales duobus rectis; sed ea animalibus tribuit quatenus viventia sunt, proinde universam viventis naturam in animalibus contemplatur; quemadmodum medici, qui humani corporis anatomen faciunt, hunc quidem hominem inspiciunt, sed ea, quae inspiciunt, ut hominis considerant, non ut huius hominis. Movit autem Aristotelem ad hoc faciendum ipsa doctrinae facilitas, quandoquidem multa sunt animalibus et plantis communia, quae facilius in animalibus cognoscuntur, in stirpibus difficilius, ideò in animalibus fuere prius declaranda, ut postea in stirpibus facilius noscerentur.°

16 Talis autem de animalibus tractatio certè ordinatissima fuit, nam cuiusque rei cognitio optima traditur, si prius eius rei naturam atque substantiam, deinde eiusdem propria accidentia declaremus. Etenim si substantia rei ignorata ab accidentibus exordiremur, nunquàm eorum scientia potiremur, quia causas ignoraremus. Atqui substantia cuiusque naturalis corporis ex materia et forma constituitur, ideò sunt prius cognoscendae hae duae essentiales partes, deinde affectiones ab eis prodeuntes, hoc ordine usus est Aristoteles in tractatione de corpore naturali communi in libris Physicorum, prius enim eius principia materiam et formam in duobus prioribus libris, deinde eiusdem accidentia in sex posterioribus consideravit. Hoc idem in corporibus animatis faciendum fuit. Est autem animati corporis forma anima ipsa, propria verò materia sunt ipsae partes instrumentales, in quibus inest anima

But with the treatment on animals he combined the one that is about what is common. For he decided to consider in animals both those [qualities] that are proper to animals and those that they have in common with plants. [He did this] not because he attributes these common things to animals as they are animals — for that would be the same as to demonstrate that an equilateral triangle has three angles equal in total magnitude to two right [angles] — but he ascribed these to animals insofar as they are living things. Thus the universal and whole nature of living [thing] is contemplated in animals, just as physicians who perform an anatomy of the human body observe, of course, this man, but consider the things that they observe as pertaining to man [as such], not as pertaining to *this* man. And ease itself of teaching moved Aristotle to do this, since there are many things common to animals and plants that are known more easily in animals, but with more difficulty in plants. Because of this, they had to be made clear first in animals so that afterward they could become known° more easily in plants.

Now certainly such a treatment on animals was most orderly, 16 for optimal knowledge of each thing is conveyed if we first make clear the nature and substance of the thing and then its proper accidents. Indeed if we began from accidents, with the substance of the thing unknown, we would never get hold of scientific knowledge of them, because we would be ignorant of causes. Yet the substance of any natural body is constituted from matter and form, and so first these two essential parts and then the affections issuing from them have to be known. Aristotle used this order in the treatment on natural body in general in the books of the *Physics*. For first, in the two earlier books, he considered its beginning-principles, matter and form, and then in the six later [books] its accidents. The same thing had to be done with animate bodies. And the form of an animate body is the soul itself, and the instrumental parts themselves are its proper matter; to them, as to

93

tanquàm in materia, et quibus utitur ad edendas operationes vitae. De hac materia prius agendum fuit, quàm de forma, dirigitur enim materia ad receptionem formae et ad formae cognitionem necessaria est, quia vires ac facultates animae non benè cognosci possunt sine cognitione partium instrumentalium, in quibus exercentur.

17   Caeterùm dum corpus animatum[36] commune consideramus, de ipsius quidem forma loqui possumus, ea enim est anima nutritiva, qua et animalia et stirpes participant, et eaedem in his omnibus sunt huius animae facultates, nutritiva, accretiva et generativa. Attamen de materia nihil commune dicere possumus, quia non eisdem instrumentis in animali et in stirpe haec anima utitur. Cùm enim et animalis temperaturam et partes ipsas non modò ad dicta munera, sed ad sentiendum quoque natura direxerit,[37] aliud in animalis corpore, aliud in stirpe organorum genus esse debuit. Nulla igitur tractatio de communi materia potuit anteponi. Propterea Aristoteles à propria animalium materia exorsus est, quae sunt ipsae animalium partes, de quibus prius pingui Minerva et sine causarum traditione loqui voluit in libris de Historia animalium, deinde accuratius in libris de Partibus animalium, ut à facilioribus ad difficiliora progrederetur. Cognita materia se contulit ad considerationem formae in libris de Anima, in qua tractatione ordinem ab universalibus ad particularia optimè servavit. Quoniam enim aliqua considerari poterat communis anima stirpium et animalium, ab hac voluit exordiri, deinde de propria animalium anima, quae sensitiva dicitur, tandem de hominis anima sermonem fecit, ad harum omnium animae partium notitiam° satis fuit cognitio materiae tradita in libris de Animalium partibus, in his enim omnes dictarum partium animae operationes exercentur.

matter, the soul belongs, and it uses them to bring about the activities of life. So matter had to be dealt with prior to form, for matter is directed toward the reception of form and is necessary for knowledge of the form, because the powers and faculties of the soul cannot be well known without knowledge of the instrumental parts in which they are exercised.

Now when we consider animate body in general, we can, of course, speak about its form, for that is the nutritive soul, in which both animals and plants participate; and in all these [i.e., in both plants and animals] the nutritive, augmentative, and generative faculties of this soul are the same.[57] But we can say nothing general about the matter, because this soul does not use the same instruments in animal and in plant. For since nature directed both the balance of an animal and the parts themselves not only to the jobs mentioned but also to being sentient, there ought to be one kind of organs in the body of an animal and another in a plant. No treatment on matter in general, therefore, could be placed first. And so Aristotle began with matter proper to animals, which is the parts themselves of animals. He wanted to speak about these first just coarsely and without the conveying of causes in the books *On the History of Animals*, and then more precisely in the books *On the Parts of Animals*, so that he could progress from easier things to more difficult ones. Matter now being known, he devoted himself to consideration of form, in the books *On the Soul*, in which treatment he maintained order in the best way, from universals to particulars. For since the soul common to plants and animals could be considered to some extent, he wanted to begin with that. Then he discoursed about the soul proper to animals, which is called sensitive, and then finally about the soul of man. The knowledge of matter conveyed in the books *On the Parts of Animals* was enough for knowledge° of all these parts of the soul, for in these are exercised all the activities of the said parts of the soul.

17

18   Poterat autem Aristoteles cum tractatione de Anima vegetali tractationem coniungere de communibus accidentibus animati corporis, deinde ad propriam animalium animam, eorumque propria accidentia transire. Maluit tamen ob doctrinae facilitatem continuata oratione de omnibus animae partibus in uno ac eodem libro disserere absque ulla tractatione de accidentibus, propter similem earum omnium naturam et essendi modum, quod etiam ab Avicenna notatur in prooemio sexti Naturalium,[38] est enim omnis anima actus primus corporis naturalis instrumentalis. Quòd si in tractatione de Anima videatur esse intermista consideratio aliquorum accidentium et operationum animae, ignorare non debemus haec non propter se tractari, sed propter cognitionem partium et facultatum ipsius animae, cuius gratia ibi considerantur. Per se autem de his agit postea Aristoteles in libris illis, qui Parvi naturales vocantur, quemadmodum ipse testatur in primis verbis libri de Sensu et sensilibus, ubi asserit se de sola anima sermonem fecisse, de accidentibus autem proponit in aliis sequentibus libris tractandum.

19   Dum igitur tractationem spectamus de ipsa viventium corporum substantia, ordo ab universalibus ad particularia non est perturbatus, cùm enim cognitio essentiae cuiusque rei magis in cognitione formae consistat, quàm in cognitione materiae, non possumus negare Aristotelem prius declarasse substantiam viventis corporis, quàm substantiam animalis. Id solùm obiici potest, quòd materiam particularem praemiserit formae universali, sed hoc melioris doctrinae gratia factum est, non poterat enim forma benè cognosci sine cognitione materiae, praesertim cùm ea materia sit corpus actu existens et sensile, proinde multò notius forma. Quoniam igitur nulla tractatio de communi materia praemitti poterat, quia nullae sunt instrumentales partes, quas easdem habeant animalia

Now Aristotle could have conjoined the treatment on the com-   18
mon accidents of animated body with the treatment on the vegetative soul and then passed over the soul proper to animals and their
proper accidents. Nevertheless, in continuing [his] speech on all
the parts of the soul, he preferred to continue his discussion in
one and the same book, without any treatment of accidents, because of the ease of teaching and on account of the similar nature
and mode of being of them all. (This is noted by Avicenna in the
proem to the sixth book of the *Naturalia*.)[58] For every soul is the
first actuality of a natural, instrumental body.[59] And so, if in the
tract *On the Soul*, consideration of some accidents and activities of
the soul appear to be intermingled, we ought not to ignore that
these things are treated not for their own sake, but for the sake of
knowledge of the parts and faculties of the soul itself, on account
of which they are considered there. Aristotle deals with them *per se*
afterward, in the books that are called the *Parva Naturalia*, just as
he attests in the first words of the book *On Sense and Sensibles*,[60]
where he asserts that he has discoursed about the soul alone and
he proposes to treat accidents in other subsequent books.

When we look, therefore, at the treatment on the substance it-   19
self of living bodies, the order from universals to particulars is not
disturbed. For since knowledge of the essence of each thing consists more in knowledge of form than in knowledge of matter, we
cannot deny that Aristotle made the substance of living body clear
prior to the substance of animal. This one objection could be
raised: he placed the particular matter before the universal form.
But this was done for the sake of better teaching. For the form
could not be known well without knowledge of the matter, especially since the matter is the actually existing and sensible body,
and accordingly much more known than form. Since no treatment
on matter in general, therefore, could have been placed first, because there are no instrumental parts that animals and plants have

et stirpes, nisi proportione quadam, ideò censuit Aristoteles melius esse aliquam considerare materiam, quàm nullam,[39] cùm etiam propria animalium materia satis conferat ad cognitionem omnis animae adipiscendam.

20   Quod[40] verò de parvis naturalibus dicebatur, levis momenti est, in illis enim de accidentibus viventium corporum agitur, ordo verò ab universalibus ad particularia potissimùm propter essentiae cognitionem solet esse necessarius, nisi enim cognitio substantiae hominis cognitam postularet substantiam animalis, possent ignoratis accidentibus animalis accidentia hominis perfectè cognosci, nisi eorum species essent. Igitur pari ratione dum et animalis et hominis natura cognita est, nulla nos necessitas cogit ut prius accidentia animalis, quàm accidentia hominis demonstremus. Sed vel doctrinae facilitas attenditur, vel si haec non urgeat, aliud quippiam respicitur; quemadmodum in illis parvis naturalibus cernere possumus.

21   Nam si Aristoteles totam de corpore vivente latè sumpto tractationem seorsum à tractatione de animalibus et de stirpibus absolvere voluisset, in ea certè omnes quoque communes operationes et affectiones viventium tractari oportuisset ante omnes de animalibus libros. Sed cùm decreverit in eodem libro continuata oratione agere de forma et viventis et animalis et hominis nulla de accidentibus tractatione interposita, manserunt postea considerandae affectiones omnes tum viventis tum animalium propriae, ut ipse testatur in principio libri de Sensu et sensilibus. De his igitur omnibus tractationem aggrediens post cognitam naturam et viventis et animalis, nulla certè necessitate coactus fuit ad prius declaranda viventis accidentia, quàm animalis, cùm neque species illorum haec sint.

as the same, except in some sort of comparative relation, Aristotle deemed it better to consider some matter rather than none, since even the matter proper to animals contributes enough to obtaining knowledge of every soul.

Now what was said about the *Parva Naturalia* is of light moment. In them the accidents of living bodies are dealt with; and especially for knowledge of essence, the order from universals to particulars is normally necessary. For unless knowledge of the substance of man demanded that the substance of animal be known [first], the accidents of man, unless they were species of the accidents of animal, could be perfectly known, even with the accidents of animal unknown. By the same reasoning, therefore, when the nature of both animal and man is known, no necessity forces us to demonstrate the accidents of animal prior to the accidents of man. Instead, either ease of teaching is observed or, if this does not urge [us], something else is regarded, as we can discern in the *Parva Naturalia*.

For if Aristotle had wanted to complete a whole treatment on living body, taken in a broad sense, separately from a treatment on animals and plants, he would certainly have needed to treat in it also the common activities and affections of living things, before all the books on animals. But since he decided, in continuing [his] speech, to deal in the same book with the form of living thing, animal, and man, with no treatment on accidents interposed, all the proper affections, both of the living and of animals, remained to be considered afterward, as he attests in the beginning of the book *On Sense and Sensibles*.[61] Undertaking a treatment on all these [subjects], therefore, after the nature of both living thing and animal were known, he was certainly not forced by any necessity to make clear the accidents of living thing prior to [those] of animal, since the latter are not species of the former.

22   Restabat igitur ut facilitatem cognitionis spectaret. Sed neque
haec ibi locum habuit, quia neque communium accidentium cognitio confert ad propriorum notitiam° facilius assequendam, neque
haec ad illam. Propria quidem animalium est tractatio de Sensu et
sensilibus, de memoria et reminiscentia, de somno et vigilia, de
motu animalium, ac de ipsorum generatione; communis autem
omnibus viventibus ea, quae est de iuventute et senectute, de longitudine et brevitate vitae, de vita et morte, de sanitate et morbo.
At quomodo horum cognitio faciliorem reddat illorum cognitionem, aut è converso illa ad haec conferant, vix imaginari aliquis
potest.

23   Aristoteles igitur neque necessitate neque facilitate ulla doctrinae coactus aliud quiddam in iis libris disponendis observare voluit et proprias animalium affectiones communibus anteponere
constituit, ut ipse videtur polliceri in primis verbis libelli de Sensu
et sensilibus, quae sunt haec, 'Postquàm de anima dictum est secundùm seipsam, dicendum est de animalibus et vitam habentibus
omnibus, quae sint propriae, et quae communes operationes ipsorum,' prius enim nominat animalia, quàm viventia omnia et prius
proprias operationes, quàm communes. Illius autem ordinis ratio
haec fuit, iam diximus Aristotelem constituisse in animalibus declarare tum accidentia propria animalium tum communia viventibus omnibus, quoniam igitur in animalibus haec omnia contem-
plabatur, ideò in iis declarandis eum ordinem servare voluit, quem
habent respectu naturae ipsius animalis.

24   Itaque illa, quae animali magis essentialia erant, prius considerare voluit, quae verò minùs essentialia, posterius. Maximè quidem
omnium essentialis est animali sensus, siquidem necessarius ei est
ut sit, animal enim per sensum est animal. Post sensum motus,
hic enim animalibus, quae ex se moventur, est necessarius non ut

It remained, therefore, that he looked to ease of knowledge. But   22
this did not have a place there either, because knowledge of common accidents does not contribute to the easier securing of knowledge° of proper ones, nor the latter to the former. Of course, the tracts *On Sense and Sensibles, On Memory and Remembrance, On Sleep and Wakefulness, Movement of Animals,* and *Generation of Animals* are proper to animals, and those common to all living things are *On Youth and Old Age, On Length and Shortness of Life, On Life and Death,* and *On Health and Disease.*[62] But in what way knowledge of the latter renders knowledge of the former easier, or conversely [in what way] the former contribute to the latter, one can hardly imagine.

Aristotle therefore, forced neither by necessity nor by any ease   23
of teaching, wanted to observe something else in disposing those books and decided to place the proper affections of animals before the common, as he appeared to promise in the first words of the little book *On Sense and Sensibles,*[63] which are these: "Now that the soul has been talked about just in and of itself, animals and everything having life—which activities are proper to them and which common—have to be talked about." He names animals prior to all living things and proper activities prior to common [ones]. This was the reason for that order: We just said that in animals Aristotle decided to make clear both the accidents proper to animal and [those] common to all living things. Since therefore all these things were contemplated in animals, in making them clear he wanted to maintain the order that [the accidents] have with regard to the nature itself of animal.

Thus, he wanted to consider first those things that were more   24
essential to animal and then afterward those less essential. And most essential of all to animal, of course, is sense, since indeed that is necessary to it as such, for an animal is an animal because of sense. After sense, motion: for this, in animals that are moved by themselves, is not necessary for them as such, but so that life

sint, sed ut servari usque ad perfectam aetatem possint, nam si
homo vel equus absque pedibus nasceretur, esset quidem aliquan-
diù homo vel equus etiam sine pedibus, sed cùm victum sibi quaerere non posset,[41] brevi interiret, nemo enim ipsi cibum afferret, si
omnis essent huiusmodi. Succedit motui generatio, quae neque
necessaria est animali ut sit, neque ut individuum ipsum servetur,
sed solùm ut servetur species, nam si omnibus hominibus, qui
nunc vivunt, auferretur vis generandi sibi simile, illi adhuc homines essent, et eorum plurimi possent diutissimè vivere, sed
tamen humana species post eorum interitum deleretur. Aristoteles
igitur primo loco de sensu et sensilibus agere voluit, et de memoria, et aliis operationibus ad sentientem facultatem attinentibus;
postea de motu animalium; deinde verò de generatione. Adde
quòd generatio simpliciter sumpta est operatio animae vegetalis,
licèt à propria animalis natura ad hanc speciem restringatur, quae
dicitur animalium generatio, ideò remotior est ab ipsa essentia
animalis, quàm sensus et motus.

25   Post generationem animalium consideravit Aristoteles operationes communes omnibus viventibus, non quidem ut ad animalia
restrictas, sic enim essent animalium propriae, sed ut communes,
quae adhuc remotiores sunt ab animalis natura, cùm non hanc,
sed aliam communiorem naturam, quae in animali inest, conse-
quantur. Hae autem sunt iuventus, senectus, longitudo et brevitas
vitae, vita et mors, sanitas et morbus. Inter has tamen posuit Aristoteles tractationem de Respiratione, licèt animalium propria sit,
eorumque non omnium, facilioris doctrinae gratia, ut ipse in eo
loco testatur, dicit enim respiratione vitam animalium conservari,
et sine hac ne modicum quidem tempus durare posse, proinde
coniungendam esse tractationem de respiratione cum tractatione
de vita et morte, tamdiù enim animalia vivunt, quamdiù possunt
respirare.

can be preserved all the way to its completion. For if a man or horse were born without feet, it would, of course, still for a time be a man or a horse, even without feet. But since it could not seek after provisions for itself, it would soon pass away. For if all were of this type, no one would bring it food. Generation follows motion, which is not necessary for an animal as such, or so that the individual is preserved, but only so that the species is preserved. For if all men who are now living lost the power to generate something similar to themselves, they would still be men, and many of them could live a very long time, but then after their death the human species would be obliterated. Aristotle therefore wanted in the first place to deal with sense and sensibles, and memory, and the other activities pertinent to the sensing faculty, and afterward with the motion of animals, and then with generation. Add [to this] that generation taken absolutely is an activity of the vegetative soul, granted that it is restricted down from the proper nature of animal to this species, which is called the generation of animals. Because of this, it is more remote from the very essence of animal than are sense and motion.

After the generation of animals, Aristotle considered activities   25 common to all living things, not, of course, as restricted to animals, for these would be proper to animals, but as common; these are still more remote from the nature of animal, since they do not ensue from it, but from another more common nature that belongs to animal. These are youth, old age, length and shortness of life, life and death, health and disease. Amid these Aristotle placed the tract *On Respiration* — granted that it is proper to animals and not to all [living things] — for the sake of easier teaching, as he attests in that passage. For he says that the life of animals is conserved by respiration, and, of course, cannot last even a short time without it. Accordingly the tract on respiration has to be conjoined with the tract on life and death.[64] For animals live as long as they can breathe.

26 Sed nimium fortasse digressi sumus. In his autem et Euclidis et Aristotelis de animalibus libris seriem tractationum considerare voluimus non ut res naturales vel mathematicas doceremus, sed ut haec tanquàm exempla ob oculos studiosorum ponendo, viam ac modum eis tradamus, quo alias[42] omnes eiusmodi difficultates facilè solvere et rationem omnis ordinis à quolibet probato authore servati intelligere queant.[43] À nostra enim meliore vel faciliore cognitione rationem sumentes nunquàm errabunt. Ubi verò neutra ratio attendi potest, liberum esse intelligent disponentis arbitrium, licèt hoc rarum admodum sit et non in totius disciplinae dispositione, sed in aliqua tantùm eius parte locum habeat.

But perhaps we have digressed too much. In all this, we wanted   26
to consider the sequence of treatments both in the books of Euclid
and in those of Aristotle on animals, not so as to teach natural or
mathematical issues, but so that in positing them as examples before the eyes of the studious, we may convey the way and mode by
which they can easily do away with all other problems of this type
and can understand the reason for every order maintained by any
proven author. For those taking the reason from our better or
easier knowledge will never err. And where neither reason can be
observed, they will understand that the choice of the one doing
the disposing is unconstrained, even granted that this is very rare
and has a place not in the disposition of the whole discipline, but
only in some part of it.

LIBER SECUNDUS

: I :

De speciebus ordinis aliorum sententia.

1 Postquàm de Ordine in universum disservimus, ad eius species descendendum est, in quibus duplex nobis offertur difficultas, una quidem de ipsarum specierum numero, altera verò de differentiis, per quas ordo in eas species dividatur. Si utramque dissolverimus, nota simul[1] fient singulorum ordinum propria officia propriaeque naturae. Nam cognitis differentiis, per quas genus in species diditur, specierum quoque definitiones manifestae fiunt.

2 Arduum nobis certamen subeundum est cum iis omnibus, qui hactenus de hac re scripserunt, praesertim cum medicis, qui magistri sui Galeni sententiam declarare ac defendere voluerunt. Galenus enim in principio libri de Arte medicinali tres ordines ponit, compositivum, resolutivum ac definitivum, quam sententiam omnes medici ad unum secuti sunt, ita enim iurarunt in verba magistri, ut solius Galeni authoritate freti nullam illius dogmatis rationem investigaverint, quasi nefas esse arbitrantes tanti viri sententiam in dubium revocare et eius rationem expetere. Propterea multos vidi eam sententiam profitentes, ipsam tamen non satis intelligentes, ex quibus si quis rationem exquireret, nihil, quod dicerent, haberent.

3 Nos verò Galeni quidem authoritatem, ut par est, magni faciamus, sed tamen rationem ne contemnamus, eamque Galeno praeferre ne vereamur, si ab ipsa Galenum dissentire invenerimus.

BOOK TWO

The position of others regarding the species of order.

Now that we have discussed Order as a universal whole, we have 1 to descend to its species. In these, a problem of two types is presented to us. One is about the number of the species themselves, and the other about the differentiae by means of which order is divided into those species. If we resolve both, each order's own proper functions and proper natures will at the same time become known. For when the differentiae by means of which a genus is divided into species become known, the definitions of the species become manifest also.

Upon us has come a tough dispute with all those who have 2 written on this matter up until now—especially with the physicians, those who wanted to defend and make clear the position of their master, Galen. For Galen posits three orders at the beginning of the book *The Art of Medicine:*[1] compositive, resolutive, and definitive. All the physicians were at one in following this position. They swore such allegiance to the master's words that, relying on the authority of Galen alone, they investigated no reason for his doctrine, as if thinking it is wicked to call into doubt the position of such a man and to request a reason for it. And so I saw many professing this position but not sufficiently understanding it. If anyone had asked them for a reason, they would have had nothing to say.

We too, of course, uphold the authority of the great Galen, as 3 is right, but nevertheless we would not contemn reason; nor would we be afraid to prefer it to Galen if we came to disagree with

Non solum enim si scopum attigerimus, aut propè ad veritatem accesserimus, verùm etiam si aliquantum ab ea nos aberrare contigerit, nihilominus laudandi erimus. Praestat enim veritatis amore ductos in aliquem errorem incidere, quàm Galeno addictos in sola ipsius authoritate acquiescere sine rationis indagatione. Illud namque ingenuum ac philosophicum animum prae se fert, hoc verò servilem.

: II :

De differentiis ordinem doctrinae dividentibus
secundùm aliorum opinionem.

1 Omnes autem[2] medici, qui hoc Galeni dictum consideraverunt, illud verum esse statuentes, ac fermè pro axiomate habentes, id tantùm declarare aggressi sunt quibus differentiis hi tres ordines à se invicem discrepent, eas enim differentias, ac totam ipsorum contemplationem ad Galeni sententiam accommodare maluerunt, quàm veritatem liberè investigando cum ea Galeni dictum conferre et videre an consonet.

2 Quod autem plurimi hac in re dicere solent, hoc est, ordinem compositivum esse illum, qui à primis principiis et à simplicissimis in eo genere incipiens transit ad composita, quae ex illis principiis producuntur. Resolutivum verò, qui ab ultimo et à fine et à composito exordiens pergit ad simpliciora, donec ad prima ac simplicissima principia perveniat. Quocirca contrarius videtur ordo resolutivus compositivo. Denique definitivum esse illum, qui neque à primo neque ab ultimo sed à definitione auspicatur, cuius singulas

Galen because of it. For not only if we were to reach the goal or come near to the truth, but even if it happened that we wandered a little bit from it, we would nonetheless deserve to be praised. For it is better to fall into error, led by love of truth, than, committed to Galen, to acquiesce in his authority alone without tracking down a reason. For the former exhibits a noble and philosophical soul, but the latter a servile one.

: II :

On the differentiae dividing order of teaching, according to the opinion of others.

Now all the physicians who considered this dictum of Galen's, 1 deciding it is true and holding it nearly as an axiom, undertook to clarify it only by what differentiae the three orders are different from each other. For they preferred to apply these differentiae and the whole contemplation of them to Galen's position, rather than, by freely investigating the truth, to compare Galen's dictum with the truth and see whether it is consonant.

And what many normally say on this issue is this: Composi- 2 tive order is that which, starting from first beginning-principles and the simplest in the genus, passes to the composites that are produced from those beginning-principles. And resolutive is that which, beginning from the ultimate, from the end, and from the composite, goes directly on to the simpler until it arrives at the first and simplest beginning-principles. Therefore resolutive order appears to be the contrary of compositive. Lastly, definitive is that which commences neither with the first nor with the ulti- mate but from the definition. By treating each of the definition's

partes tractando, donec omnes absolvantur servari dicitur ordo definitivus, qui propterea diversus apparet à compositivo et à resolutivo.

3 Verùm aliqui diligentius haec omnia perpendentes et hanc sententiam magis declarare volentes dicunt ex definitione ordinis doctrinae colligi id esse officium ordinis, ut omnia per ipsum convenienter disponantur, rectam autem dispositionem esse quando ad unum omnia referantur et ab uno pendeant. Cùm igitur illud unum possit esse triplex, tres ordinis species oriuntur, vel enim est unum principium vel unus finis vel unum medium, à quo cuiusque disciplinae exordium sumitur et ad quod alia omnia referuntur. Unum quidem principium si fuerit, à quo sumatur exordium et à quo tota scientia pendere dicatur, oritur ordo compositivus. Si verò unus finis sit,[3] ordo resolutivus, qui ab ultimo fine auspicatur et transit ad ea omnia investiganda, quae ad finem ullum[4] ducere possint, illa semper ad ipsum finem prius cognitum referendo. Si verò sit unum medium, est ordo definitivus, hic enim cùm à definitione, quae totam artem complectitur, exordiatur, videtur quodammodo auspicari à centro, à quo omnes lineae ad circumferentiam ductae originem trahunt, nam progressus ad singulorum considerationem, donec omnes eius definitionis partes tractatae fuerint, est veluti à centro ad circumferentiam per omnes lineas transire et quemadmodum lineae omnes modo quodam in centro contineri dicuntur, ita etiam tota ars in illa definitione, à qua ipsius exordium sumitur. Ordo igitur definitivus ab una definitione tanquàm ab uno medio auspicatur, ad quam reliqua omnia referuntur et à qua pendent.

4 Itaque sufficiens est ordinis partitio in has tres species, quia cùm ordinata doctrina illa sit, quae ab uno incipit, illud unum non

parts, until all have been dealt with, definitive order is said to be maintained. And so it appears different from compositive and from resolutive.

But some, examining all this more carefully and wanting to 3
make this position more clear, say that from the definition of the order of teaching the function of order is gathered to be as follows: that by means of it all things are appropriately disposed and, furthermore, that the correct disposition is when all things are referred to one thing and depend on one thing. Since, therefore, that one thing can be of three types, three species of order arise — for it is either one beginning-principle or one end or one middle from which the beginning of any discipline is taken and to which everything else is referred. Now if there was one beginning-principle from which the beginning [of the discipline] is taken and on which the whole science may be said to depend, then compositive order arises. And if there is one end, then [there is] resolutive order, which commences with the ultimate end and passes to all the things being investigated that can lead to that end, by referring those things always to that earlier known end itself. And if there is one middle, then there is definitive order. This, since it begins from a definition that encompasses the whole art, appears in a sense to commence with a center from which all lines leading to the circumference draw their origin. For the progression to the consideration of each [part], until all the parts of its definition have been treated, is just as to pass through all the lines from the center to the circumference. And just as all lines are said in some way to be contained in the center, so too is the whole art in that definition from which its beginning is taken. Definitive order, therefore, commences with one definition, as from one middle, to which all the rest are referred and on which they depend.

And so the partition of order into these three species is suffi- 4
cient, because, since ordered teaching is that which starts from

potest esse nisi unum principium vel unum medium vel unus finis. Quare nec plures ordinis species dantur, nec pauciores.

: III :

In quo dictae differentiae confutantur et ostenditur,
non esse ordinis conditionem ab uno incipere.

1 Haec omnia, quae ab his dicuntur, pulchra quidem sunt et ingeniosè excogitata, at si diligenter expendantur, plurimum in se difficultatis habent, cùm ab eis multa gratis et sine ulla ratione dicantur, multa etiam, quae apertè repugnant veritati.

2 Primùm quidem illud considerandum est, quod dicunt ordinatam esse doctrinam si ab uno omnia pendeant, à quo tractationis exordium sumatur. Huius quidem dicti nullam ipsi rationem attulerunt, cùm per eorum verba non appareat cur, si à pluribus initium doctrinae sumatur, ordinata ea doctrina appellari non possit. At certè si res ita se haberet, in eo saltem essent reprehendendi, quòd talem conditionem in definitione ordinis doctrinae non expresserunt, non fuit enim praetermittenda, si est essentialis et necessaria ordinis conditio, sed definitio debuit esse talis, Ordo doctrinae est habitus instrumentalis, quo docemur totam disciplinam ita disponere, ut ab uno initium tractationis sumatur, ad quod reliqua omnia referantur. Sic enim expressa esset in definitione rectitudo illa dispositionis, quae ipsam ordinis essentiam constituit,

one thing, that one thing can be nothing except one beginning-principle or one middle or one end. Therefore there are no more species of order, and no fewer.

: III :

In which the said differentiae are confuted, and it is shown that to start from one thing is not a characteristic of order.

All these things that are said by them are, of course, beautiful and 1
ingeniously imagined. But if they are weighed carefully [it will be seen that] they have in them plenty of problems, since many of them are said gratuitously and without any reason; in reality there are many that are plainly incompatible with the truth.

First, of course, it has to be considered that they say that teach- 2
ing is ordered if all things depend on one thing from which the beginning of the treatment is taken. But, of course, they brought forward no reason for this dictum itself, and it is not apparent from their words why, if the start of teaching were taken from many things, the teaching could not be called ordered. And certainly if such had been the case, they would have to be censured in this at least, that they did not expressly state such a characteristic in the definition of the order of teaching. For it should not have been overlooked if it is an essential and necessary characteristic of order. Instead the definition ought to have been like this: the order of teaching is an instrumental habit by which we are taught to dispose a whole discipline, such that the start of a treatment is taken from one thing, to which all the rest are referred. For thus in the definition there would have been expressly stated that correctness of disposition which constitutes the very essence of order.

unde postea differentias illas, quibus ordinem in species partiuntur, haurire potuissent.

3   Verumtamen ea conditio in ordine doctrinae non requiritur, sed
est ei accidentalis, in ordine enim compositivo non est necessarium, ut ab uno principio exordium scientiae sumatur, cùm à pluribus sumi queat, ut patet in scientia naturali ab Aristotele tradita
ordine compositivo, incipit enim non ab uno rerum naturalium
principio, sed à pluribus, et res ipsae naturales non ad unum principium sed ad plura relationem habent. Ideò Aristoteles in primo
Physicorum libro tractationem de principiis in plurali numero
proponit, non in singulari. Similiter in toto illo primo libro et in
prooemio primi Meteorologicorum prima principia nominat in
numero plurali. In primo quoque libro de Moribus, capite quarto,
ordinem compositivum significare volens non dicit à principio, sed
à principiis. Huius autem ratio sumitur ex iis, quae dicuntur ab
Aristotele paulò ante finem primi libri Posteriorum Analyticorum,
ibi namque docet unitatem scientiae ex unitate subiecti pendere,
non ex unitate primi principii, imò dicit affectionum vel principiorum vel specierum multitudinem non impedire scientiae unitatem,
dummodò unum sit genus subiectum, à quo tanquàm à communi
radice omnia prodeant. Non est igitur necessarium, ut scientia
auspicetur ab uno principio, cùm unius subiecti plura prima principia esse possint.

4   Haec autem subiecti unitas nullum ordinem constituit, neque
compositivum neque resolutivum neque aliquem alium, quia subiectum totam scientiam capit, unde etiam solet vocari adaequa-
tum, non est igitur magis primum, quàm ultimum, nec scientia
ab eo auspicari dicitur, cùm in nulla scientiae parte quaeratur,
adsit tamen omnibus partibus, quatenus omnibus, quae in scientia

Then from there they could afterward have derived those differentiae by which they divide order into species.

But this characteristic, however, is not required in the order of 3 teaching but is accidental to it. For in compositive order, it is not necessary that the beginning of a science be taken from one beginning-principle, since it can be taken from many, as is patent in natural science conveyed by Aristotle using compositive order. For he begins not from one beginning-principle of natural things, but from many, and the natural things themselves are relative not to one beginning-principle but to many. Because of this, Aristotle in the first book of the *Physics*[2] set out the treatment on beginning-principles in the plural, not in the singular. Similarly in the whole of that first book and in the proem of the first [book] of *On Meteorology*,[3] he names first beginning-principles in the plural. And also in the first book of the *Ethics*, chapter 4,[4] wanting to indicate compositive order, he does not say from [a] beginning-principle but from beginning-principles. The reason for this is taken from the things that are said by Aristotle a little before the end of the first book of the *Posterior Analytics*.[5] For there he teaches that the unity of a science depends on unity of [its] subject, not on unity of [its] first beginning-principle. He even says a multitude of affections, beginning-principles, or species does not impede the unity of a science, as long as the subject, from which everything issues as from a common root, is one genus. It is not necessary, therefore, that a science commence with one beginning-principle, since there can be many first beginning-principles of one subject.

This unity of subject, however, establishes no order, neither 4 compositive nor resolutive nor any other, because the subject takes in the whole science. It is thus normally said to be coextended [with the science], and is, therefore, no more first than ultimate. Nor is the science said to commence with it, since it is inquired after in no part of the science. Rather, it is present in the parts, insofar as it extends under everything that is treated in the science.

tractantur, substernitur. À subiecto igitur et ab eius unitate nullus
ordo desumitur, principium autem unum vel unum accidens subiecti esse non est necessarium, cùm possint esse plura, ergo non
datur aliquod unum in scientia, à quo exordium sumendo compositivus ordo constituatur.

5   Illud quoque non rectè dicitur, omnia quae in scientia considerantur, relationem habere ad unum primum principium, dato
enim quòd huiusmodi principium primum in scientia unum sit,
non tamen verum est quòd aliorum omnium eatenus consideratio
habeatur, quatenus ad illud referuntur, principia enim subiecti
considerantur cum relatione ad ipsum et propter ipsum, non è
converso subiectum propter ipsa, neque cum relatione ad illa; quia
principia subiecti quatenus principia sunt, respectum habent ad id,
cuius sunt principia, ut principia corporis naturalis ad ipsum corpus naturale. At subiectum ipsum nullam habet relationem ad
principia, sed omni respectu° absolutum° est.

6   Quae igitur ab illis dicuntur, nullam in ordine compositivo veritatem habent; videntur autem magis competere resolutivo, hic
enim non modò à fine, sed ab uno fine tractationis exordium
sumit. Attamen non ideò ab uno fine auspicatur, quia ordo et recta
partium dispositio id requirat, sed quia artis unitas postulat unitatem finis, nam unitas finis ad ordinem ac dispositionem nihil
confert, sed ad artis unitatem, ordo enim resolutivus potest tam[5]
ab uno, quàm à pluribus finibus incipere, attamen si à pluribus,
plures erunt artes, non una. Est igitur ordinis resolutivi conditio
auspicari ab uno fine, tamen non quatenus uno, sed quatenus fine.

7   Ab uno igitur incipere non est essentialis conditio ipsius ordinis, cùm in ordine compositivo minimè vera sit. In resolutivo autem locum quidem habeat, non tamen per se, sed ex accidenti,

No order, therefore, is drawn from the subject and from its unity. And so it is not necessary that there be one beginning-principle or one accident to a subject, since there can be many. Compositive order, therefore, is not established just by taking a beginning from any one thing in a science.

It is also not correct to say that everything that is considered in 5 a science is relative to one first beginning-principle. Even when there is one first beginning-principle of this type in a science, still it is not true that all the others are to be considered only insofar as they are referred to it. For the beginning-principles of a sub-ject are considered in relation to it and for the sake of it, and not, vice versa, the subject for the sake of them or in relation to them, because beginning-principles of a subject, insofar as they are beginning-principles, have a relationship to that of which they are the beginning-principles, just as the beginning-principles of natural body have to natural body itself. But the subject itself is not relative to the beginning-principles; it is in every respect° absolute.°6

The things said by them, therefore, have no truth in composi- 6 tive order; they appear to apply more to resolutive. For the latter takes the beginning of a treatment not only from an end, but from one end. But nevertheless, it does not so commence with one end because the order and correct disposition of the parts require it, but because unity of the art demands unity of the end. For unity of the end contributes nothing to order and disposition but [does contribute something] to unity of the art. For resolutive order can start as much from one as from many ends. But if from many, there are many arts, not one. It is a characteristic of resolutive or-der, therefore, to commence with one end, nevertheless not insofar as it is one, but insofar as it is an end.

To start from one, therefore, is not an essential characteristic of 7 order itself, since this is not true at all in compositive order. It has a place, however, in resolutive order, though not *per se* but by

quia neque quatenus est ordo, neque quatenus est ordo resolutivus, sed solùm quatenus ars illa una esse debet.

8   Cùm igitur haec non sit essentialis conditio ordinis, sed accidens idque separabile, manifestum est eos per differentias acciden-
tales et quae constituere species nequeunt, divisionem ordinis fecisse, perinde ac si dixissent animalium alia sedere, alia currere,
alia stare. Tales itaque differentiae reiiciendae sunt.

: IV :

In quo definitivus ordo refellitur.

1   Illa quoque mihi semper maximè dubia visa sunt, quae ab his de
ordine definitivo dicuntur, is enim mihi videtur non esse admitten-
dus, licèt eum Galenus tanquàm tertium ordinem et ab aliis duobus distinctum posuerit. Quapropter ea quoque, quae ad huius
ordinis declarationem et comprobationem ab aliis dicuntur, vana
et commentitia esse puto et quae parvo negotio refelli possint.

2   Afferam igitur in medium ea omnia dubia, quae me adversus
hanc sententiam movere potuerunt, summoperè ob veritatis amo-
rem desiderans aut alios cognita° veritate in meam sententiam venire, aut me, si ipse hac in re fallor, errore meo perspecto mutare
sententiam.

3   Nam, si ratio mihi persuaserit, palinodiam canere non verebor.

4   Ante omnia, cùm ordinem definitivum eum esse dicant, qui à
definitione exordium sumit, videndum est quaenam et cuiusnam
rei sit ea definitio, ut non illa tantùm, quae ab his dicuntur, sed
alia quoque omnia, quae dicere possent, reiiciendo, omnemque

accident—because [it does so] not insofar as it is order or insofar as it is resolutive order, but only insofar as that art ought to be one.

Since, therefore, this is not an essential characteristic of order 8 but is an accident and is separable, it is manifest that they made [their] division of order by means of accidental differentiae and these cannot constitute species, any more than if they had said of animals that some sit, some run, some stand. And so such differentia have to be rejected.

: IV :

In which definitive order is refuted.

The things said by them about definitive order too always seemed 1 very doubtful to me. For it appears to me that it should not be admitted, even granting that Galen posited it as a third order, distinct from the other two. And so I hold also that those things also that are said by others to clarify and confirm this order are vain and counterfeit and can be refuted with little effort.

I will bring up, therefore, all those doubts that were able to 2 move me against this position, chiefly because of a love of truth, desiring either that others, in recognizing° the truth, come to my position, or that I myself, if I am mistaken in this issue, seeing my error, change my position.

For if reason has persuaded me, I will not be afraid to sing a 3 palinode.

Before all else, since they say it is definitive order that takes its 4 beginning from a definition, it has to be seen what this definition is and [what it is] of, so that by rejecting not only the things that are said by them but also everything that they could say and by

subterfugii locum de medio tollendo ostendamus nos pro huius ordinis defensione nihil invenire potuisse.

5   Non est credendum, eos quamlibet definitionem intelligere, ut quaecumque sit ea definitio, à qua exordium doctrinae sumatur, ordo ab ea nominetur definitivus. Etenim hoc admisso sequeretur solum dari ordinem definitivum et eo uno scientias et artes et disciplinas omnes traditas esse et tradi posse, omnes enim à definitione auspicantur, ut revera auspicari debent, quia praeceptum est Platonis in Phaedro ut de aliqua re dicere aggressuri eius rei definitionem in primis proponamus. Idem praecipit Aristoteles in primo libro de Moribus capite septimo et aliis in locis. Quod etiam in omnibus suis libris observavit, librum enim de Interpretatione auspicatus est à definitione nominis et verbi et orationis et enunciationis; librum Priorum Analyticorum à definitione ratiocinationis; librum Posteriorum à definitione demonstrationis; scientiam naturalem totam à definitione naturae et corporis naturalis.

6   Scientiae quoque mathematicae à definitionibus exordiuntur. Sed quid opus est in re manifesta tempus conterere? Cuncti videre possunt, omnium disciplinarum principia esse definitiones. Non igitur omnem definitionem illi, qui ordinem definitivum excogitarunt, intellexere.°

7   Quaenam igitur est ea definitio, cui hanc praerogativam tribuamus ut ab ea ordientes ordine dicamur uti definitivo? Rationi consonum videtur eam alicuius rei, quae in illa facultate tractetur, definitionem esse. Cùm enim de eo ordine nunc sermo nobis institutus sit, qui inter res in disciplina aliqua tractatas servatur, aliqua earum esse videtur, cuius definitio toti disciplinae anteposita ordinem faciat definitivum. Attamen cuiusnam rei consideratae ea definitio esse possit equidem non video, nam alicuius primi principii

getting rid of every point of subterfuge, we can show that we have been able to discover nothing in defense of this order.

It is not to be believed that they understand the definition to be 5
just any, such that whatever the definition is from which the teaching takes its beginning, the order is called definitive because of it. For indeed if this were countenanced, it would follow that there is only definitive order, and by this one all sciences, arts, and disciplines are conveyed and can be conveyed. For all these commence with a definition, as in truth they ought to commence. For it is a precept of Plato's in the *Phaedrus*[7] that, undertaking to speak about anything, we set out in the first place the definition of the thing. Aristotle counsels the same in the first book of the *Ethics*, chapter seven,[8] and in other passages. And he observed this in all his books. For he commenced the book *On Interpretation* with definitions of noun, verb, speech, and proposition;[9] the book *Prior Analytics* with the definition of ratiocination;[10] the book *Posterior [Analytics]* with the definition of demonstration;[11] and natural science as a whole with the definition of nature and natural body.[12]

Mathematical sciences, too, begin with definitions. But what 6
need is there to waste time on something manifest? Everyone can see that definitions are the beginning-principles of all disciplines. Those who imagine definitive order, therefore, did not mean° every definition.

Which definition is it then, to which we ascribe the prerogative 7
that, beginning with it, we may be said to use definitive order? It appears consonant with reason that the definition be of any thing that is treated in that branch of learning. For since the discourse is now intended by us to be about the order that is maintained among the things treated in any discipline, it appears to be any of them whose definition, placed before the whole discipline, makes the order definitive. But for my part I do not see which thing under consideration it can be the definition of. For it cannot be [the definition] of some first beginning-principle, since when we begin

ea esse non potest, quoniam à primis principiis ordientes facimus ordinem compositivum. Neque definitio subiecti scientiae universae, ea namque ex primis scientiae principiis conflata est, proinde compositivum ordinem constituit; ut in scientia naturali exordium sumitur à definitione naturae, quae est subiectum; id est, formalis ratio subiecti totius naturalis philosophiae, imò et post definitionem naturae affert Aristoteles ipsam naturalis corporis definitionem. Ea tamen scientia non dicitur scripta ordine definitivo, sed compositivo. Sed neque definitio ultimi finis, sic enim fieret ordo resolutivus, non definitivus, ordo namque resolutivus est à notione finis, at nil aliud est à notione finis incipere, quàm ab ipsius finis definitione, ut declarat Aristoteles in 23 contextu septimi libri Metaphysicorum; et ut videre possumus in omnibus disciplinis traditis ordine resolutivo.

8   Relinquitur ut ea sit definitio alicuius medii inter prima principia et ultimum finem, quod quidem est ipsorum dictis valdè consentaneum, quandoquidem in ordine definitivo asserunt neque à primis principiis exordium sumi, is enim esset ordo compositivus; nec ab ultimo fine, esset enim resolutivus, sed à medio aliquo. Nam si ambiguè non loquuntur et eiusdem rationis apud eos est acceptio primi et ultimi et medii, medium tale, quod inter prima et ultima positum sit, intelligant necesse est, veluti si diceremus in tradenda ordine compositivo arte aedificatoria esse ordiendum à lapidibus, ab his transeundum ad fundamenta et parietes et tectum, ab his demum ad domum totam; ordine autem resolutivo incipiendum esse à definitione domus, quae est finis, inde transeundum ad fundamenta et parietes et tandem perveniendum ad prima principia et elementa, ut ad lapides. In eadem autem

with first beginning-principles we make the order compositive. Nor [can it be] the definition of the subject of the science as a universal whole, for that [definition] is assembled from the first beginning-principles of the science and so constitutes compositive order — as in natural science, the beginning is taken from the definition of nature, which is the subject, that is, the formal reason of the whole subject of natural philosophy. And indeed, after the definition of nature Aristotle brings up the definition itself of natural body. Nevertheless, that science is not said to be written using definitive order but compositive. And neither [can the definition be] the definition of the ultimate end, for then resolutive order would result, not definitive For resolutive order is from a notion of the end, but to start from a notion of the end is nothing other than [to start] from the definition of the end itself, as Aristotle makes clear in text no. 23 of the seventh book of the *Metaphysics*,[13] and as we can see in all disciplines conveyed using resolutive order.

It remains that the definition is of some middle between first beginning-principles and the ultimate end. And, of course, this agrees very much with what they say, since they assert that in definitive order the beginning is taken not from first beginning-principles, for that would be compositive order, or from the ultimate end, for that would be resolutive, but from some middle. Now if they do not speak ambiguously and they take the meaning of first, ultimate, and middle in the same way, then it is necessary that they understand middle to be such that it is located between first things and ultimate things. It is as if we were to say, in conveying the art of building using compositive order, that we have to begin with stones, go on from these to the foundations, walls, and roof, [and] from these finally to the whole house; using resolutive order, however, we have to start from a definition of the house, which is the end, and then go on to the foundations and walls and go through finally to the first beginning-principles and elements,

8

tradenda ordine definitivo esse auspicandum à definitione alicuius medii, veluti fundamentorum vel parietis.

9   Quo quidem modo si dicant ordinem definitivum exordiri à medio, eorum sententia risum potius, quàm confutationem meretur, is enim processus erit semper ab ignoto ad notum, quia necessum est omnis medii talis perfectam cognitionem pendere vel ex cognitione primorum principiorum vel ex cognitione ultimi finis. Aut igitur prima aut ultima erunt illo medio notiora et ad eius cognitionem necessaria, proinde prius tractanda, quàm ipsum medium. Ad id namque omnino confitendum ratio nos ipsa compellit, ab eorum tractatione auspicandum esse, quorum notitia° ad aliorum perfectam cognitionem acquirendam necessaria sit. Ab huiusmodi igitur medio nunquàm incipiendum erit, sed semper vel à primis principiis vel ab ultimo fine.

10   Ex his manifestum est, nullius rei consideratae esse posse definitionem illam, à qua exordiendo dicitur ordo servari definitivus.

11   Reliquum est ut dicant eam esse definitionem ipsiusmet scientiae vel disciplinae traditam ex ipsis rebus considerandis, atque haec fuit, ut mihi videtur, eorum de ordine definitivo sententia, dicunt enim hunc ordinem tunc servari, cùm in principio scientiae ipsa scientiae definitio ex omnibus tractandis rebus conflata proponitur, mox enim singulas ordinatim declarando, donec omnes tractatae sint, quae in ea definitione expressae fuerant, ordo dicitur servari definitivus, veluti si scribenda sit naturalis scientia ordine definitivo, ante omnia erit proponenda haec vel alia similis definitio scientiae illius:[6] Scientia naturalis est cognitio principiorum et accidentium omnium corporis naturalis, primùm quidem universè, deinde speciatim simplicium corporum et mistorum tum

that is, to the stones. In conveying the same [art] using definitive order, however, we have to commence with a definition of some middle, such as foundations or walls.

If, of course, they say definitive order begins from a middle in 9 this way, then their position merits laughter rather than refutation. For the proceeding will always be from unknown to known, because it is necessary that perfect knowledge of every such middle depends either on knowledge of first beginning-principles or on knowledge of the ultimate end. Either the first things or the ultimate things, therefore, will be more known than that middle and necessary for knowledge of it, and so has to be treated prior to the middle itself. Now reason itself compels us to confess altogether that we have to commence with a treatment of those things knowledge° of which is necessary for acquiring perfect knowledge of the others. Starting will always have to be, therefore, from either first beginning-principles or the ultimate end, never from a middle of this type.

From all this it is manifest that the definition from which—by 10 beginning from it—definitive order is said to be maintained, can be the definition of none of the things considered.

It remains that they could say it is the definition of the science 11 itself or of the discipline, conveyed by considering the things themselves. And that, as it appears to me, was their position on definitive order. For they say this order is maintained when the very definition of the science, assembled from all the things being treated, is set out at the beginning of the science. And then, by clarifying each thing in order, until everything that was expressly stated in the definition has been treated, definitive order is said to be maintained. Just as, if natural science were to be written using definitive order, this definition of that science, or another one similar, would be placed before all else: natural science is knowledge of all the beginning-principles and accidents of natural body—first, of course, universally, and then specifically of simple

animatorum, tum anima expertium usque ad ultimas eorum species. Hac definitione proposita incipiendum erit à tractatione primorum principiorum et communium affectionum rerum omnium naturalium, deinde de simplicibus corporibus agendum, postea de mistis, prius quidem generatim, deinde speciatim usque ad species ipsorum infimas. Sic enim absoluta erit tractatio omnium earum rerum, quae in illa definitione acceptae fuerant et scientia naturalis tradita dicetur ordine definitivo, quemadmodum his temporibus scripta ac in lucem edita est à quodam eruditissimo viro conterraneo atque amico meo, qui nisi immatura morte praereptus fuisset, magno splendori sibi ac patriae nostrae futurus erat.

12   Hoc ipso ordine scripsit Artem medicinalem Galenus, proposuit enim in primis talem artis medicae definitionem, Medicina est scientia salubrium insalubrium et neutrorum, deinde hanc definitionem in partes dissolvens, coepit agere de salubribus, mox de insalubribus, tandem de neutris. Quibus omnibus absolutis dicitur tota definitio esse declarata et ars medica tradita ordine definitivo. Propterea non incongruè aliqui talem ordinem dixerunt posse etiam in hoc sensu vocari resolutivum, quia procedit à resolutione definitionis in omnes suas partes. Dicunt etiam aliqui⁷ hunc ordinem esse veluti progressum à medio et à centro ad circumferentiam, quemadmodum enim in centro circuli sunt virtute quadam omnes lineae ab eo prodeuntes, ita in definitione artis est veluti in indivisibili ars ipsa tota, postea per singularum definitionis partium explicationem fit quodammodo transitus à medio ad circumferentiam.

13   Haec omnia si benè considerentur, pulchra potius apparebunt, quàm vera, primùm enim vana redditur ea ordinis divisio, quam aliqui fecerunt, dicentes ordinem esse, in quo sumitur exordium aut ab uno principio aut ab uno fine aut ab uno medio. Nam

bodies and of mixed, both of animate things and of things not having a soul, all the way to their ultimate species. With this definition set out, [we] will have to start with a treatment of first beginning-principles and the common affections of all natural things, and then deal with simple bodies and afterward with mixed—first, of course, generally, and then specifically all the way to their lowest species. For thus will be completed a treatment of all the things that were accepted in that definition, and it will be said that natural science is conveyed using definitive order. Such [a science] has these days been written and brought to light by a certain most learned man, my countryman and friend, who, he had not been taken away by a premature death, would have brought great honor to himself and our country.

Galen wrote *The Art of Medicine* using this very order. For in the 12 first place he set out the following definition of medical art: "Medicine is the science of the healthy, of the unhealthy, and of neither."[14] Then, resolving this definition into parts, he began to deal with the healthy, next the unhealthy and finally neither. With all this completed, the whole definition is said to be made clear and the medical art to be conveyed using definitive order. Therefore some, not inaptly, said that such an order can even be called resolutive, in the sense that it proceeds from resolution of the definition into all its parts. And some also say that this order is like a progression from the middle and from center to circumference. For just as all lines issuing from the center of a circle are in it virtually, as it were, so a whole art itself is in the definition of the art as in something indivisible. Afterward, by means of an explication of each of the parts of the definition, there is in some sense a passage from middle to circumference.

If all these things are considered well, they will appear beautiful 13 rather than true. For first, the division of order that some made, saying order is that in which the beginning is taken either from one beginning-principle or from one end or from one middle, is

aequivocè medium accipiunt, nisi in eodem sensu medium intelligant, in quo et primum et ultimum acceperunt, nempè medium tale, quod inter eiusmodi primum et ultimum sit collocatum. Cùm igitur primum et ultimum accipiantur pro rebus consideratis, medium quoque pro aliqua re considerata sumi debuit. Attamen ipsi μεταφορικῶς medium acceperunt in quadam valdè impropria et commentitia significatione, quare ipsorum divisio non est necessaria, cùm membra dividentia non possint propter ambiguitatem redigi ad contradictoria, contradictio[8] enim est oppositio omni carens ambiguitate. Dum igitur ordinem ita partiuntur, perinde faciunt ac si quis diceret omne coloratum esse aut album, aut nigrum, aut dulce, vana enim et aequivoca dici solet huiusmodi divisio et imperfecta, quia rem divisam non adimplet.

14   Praeterea si benè cognoscamus, quidnam sit definitio ipsius scientiae vel artis et quomodo ea definitio ad partes eius disciplinae referatur, nulla ratione ordinem hunc definitivum admittemus, quoniam inter definitionem alicuius scientiae et partes eiusdem nullus potest ordo considerari. Ut igitur hoc cognoscatur, scire° oportet omnem disciplinam, cui prooemium sit appositum, dividi solere in prooemium et tractationem, tractatio quidem praecipua est, imò sola est necessaria et est corpus ipsum, ut ita dicam, illius disciplinae. Prooemium verò non est ipsius disciplinae propriè dicta pars, neque est necessarium, cùm sine ipso tractatio et disciplina integra maneat, sed ab authoribus apponi solet ad lectoris docilitatem, solent enim in prooemiis proponere res considerandas et quandoque etiam ordinem, quo considerandae sunt, nihil tamen ibi consideratur, nihil tractatur, sed ubi incipit author aliquid considerare, ibi tractatio incipit, quae à prooemio distinguitur.

rendered vain. For they accept middle equivocally, unless they un-derstand middle in the same sense in which they accepted both first and ultimate, namely that middle is such that it is located between a first and an ultimate of the same type. And therefore since first and ultimate are accepted as being things under con-sideration, the middle too ought to be taken as something un-der consideration. Nevertheless, they accepted middle *metaphoricōs* (metaphorically) in some sort of highly improper and counterfeit signification. And so their division is not necessary since, on ac-count of the ambiguity, the branches in the division cannot be traced back to contradictories, for contradiction is opposition lack-ing all ambiguity. When, therefore, they partition order thus, they do so just as if someone were to say that every color is either white or black or sweet. A division of this type is normally called vain and equivocal, and also imperfect, because it does not exhaust what is being divided.

Moreover, if we knew well what the definition of a science or   14
art itself is, and in what way the definition is referred to the parts of its discipline, in no way would we admit this definitive order, since between the definition of any science and the parts of the same [science], no order can be considered. So that this may be known, therefore, it must be understood° that every discipline to which a proem is adjoined is normally divided into the proem and the treatment. The treatment, of course, is principal; indeed it alone is necessary and is the very body, as I might say it, of that discipline. The proem is not a part, properly speaking, of the dis-cipline itself, nor is it necessary, since without it, the treatment and the discipline remain intact; but it is normally adjoined by authors to make the reader more teachable.[15] For they normally set out in proems the things to be considered and sometimes also the order in which they are to be considered. Nevertheless nothing is considered there, nothing is treated; and where the author starts to consider something, there the treatment, which is distinguished

Unde patet posse à quavis disciplina auferri prooemium ea integra manente, cùm non ob necessitatem, sed propter maiorem commoditatem soleat disciplinis prooemium apponi.

15   In principio igitur alicuius scientiae si eius scientiae definitionem ex omnibus rebus considerandis constitutum afferamus, nil aliud est ea definitio, quàm prooemium, in quo res omnes tractandae proponuntur, discrimen, si quod est, est solùm in modo loquendi, nulla est enim eiusmodi definitio, quam non possimus proferre ad modum prooemii. Et è converso nullum tale prooemium, quod non possit in modum definitionis proponi. Nam ubi Galenus dicit, Medicina est scientia salubrium, insalubrium et neutrorum, eandem sententiam his verbis referre possumus, Artem medicam tradituri sumus, in qua de salubribus et insalubribus et neutris disseremus.

16   Aristoteles verò in principio primi libri Meteorologicorum prooemium facit, in quo proponit res omnes in ea scientiae naturalis parte considerandas, quam totam sententiam possumus levi negotio in illius disciplinae meteorologicae definitionem convertere dicendo; Meteorologia est scientia causas declarans flammarum omnium in sublimi apparentium et cometarum et lacteae viae et reliquorum omnium, quae ibi nominantur ab Aristotele. Quid igitur? Si per talem Meteorologiae definitionem Aristoteles eundem illius prooemii sensum retulisset, vel si nos prooemium illud in eiusmodi definitionem verteremus, factusne esset statim ordo illius scientiae definitivus, cùm in partibus tractationis et in earum serie nulla esset facta mutatio? Certè risu dignum hoc est, neque rationis capax ille est, qui hoc dicat. Simili namque ratione possemus totam scientiam naturalem ab Aristotele scriptam considerare et ne syllaba quidem tractationi ipsius adiecta vel deleta, nullaque

from the proem, starts. From this it is patent, that a proem can be removed from any discipline, while it [i.e., the discipline] remains intact, since the proem is normally adjoined to disciplines not out of necessity but on account of greater convenience.

At the beginning of any science, therefore, if we bring forward a   15
definition of that science constituted from all the things being considered, the definition is nothing other than the proem, in which all the things being treated are put forward. The discriminating difference, if there is one, is only in the manner of speaking, for no definition is of the type that we could not advance it in the manner of a proem. And conversely, no proem is such that it could not be set out in the manner of a definition. For where Galen says, "Medicine is the science of the healthy, of the unhealthy, and of neither," we can recount the same position in these words: we will convey the medical art, in which we will discuss the healthy, the unhealthy, and neither.

And Aristotle puts a proem in the beginning of the first book   16
of *On Meteorology*,[16] in which he sets out everything to be considered in that part of natural science. We can, with light effort, convert the whole position into a definition of that meteorological discipline by saying: meteorology is the science making clear the causes of all blazes appearing aloft, and of comets, and of the Milky Way, and of all the other things that are named there by Aristotle. So what? If Aristotle had recounted the same sense of the proem by means of such a definition of meteorology or if we turned that proem into a definition of that type, would the order of that science not have been made definitive straightaway although no change in the parts of the treatment or their sequence would have occurred? Certainly this is deserving of laughter, and he who says it is incapable of reason. For by similar reasoning we could consider the whole of natural science written by Aristotle, and with not even one syllable of the treatment itself added or deleted and its order changed in no part, nevertheless change the

in parte ordine illius mutato, ordinem tamen mutare et ex compositivo facere definitivum, qua quidem re nihil absurdius excogitari
potest. Scriptam enim esse eam scientiam ordine compositivo
nemo est, qui ignoret. Illi autem Aristoteles prooemium apposuit,
in quo quidem non expressit res omnes in tota scientia considerandas, sed in iis tantùm libris de Naturali auscultatione, quorum
respectum ad reliquas eius scientiae partes significare satis habuit,
nam omne prooemium est omnino arbitrarium et potest author
vel ipsum prorsus omittere vel aliqua tantùm praecipua dimissis
reliquis in eo proponere, ut ibi proposuit Aristoteles tractationem
faciendam de primis principiis, sine quorum cognitione non possunt alia benè cognosci.

17   Nos autem fingamus Aristotelem ibi plenius fecisse prooemium
et in eo res omnes in tota scientia naturali considerandas propo-
suisse hunc in modum, naturalem scientiam scripturi, de primis
principiis primo loco dicemus, quoniam absque eorum cognitione
caetera cognosci non possunt, deinde ordinatim de rebus omnibus
naturalibus disseremus, primùm quidem de simplicibus, mox de
mistis, et de his prius generatim, postea speciatim quousque ad
infimas species pervenerimus. Hoc prooemio eius scientiae constituto, in quo apparent res omnes tota illa scientia tractandae, si
tractatio tota sequens integra et immutata servetur et eadem serie,
qua eam scripsit Aristoteles, quonam ordine scripta dicetur?
Nonne compositivo? Proculdubiò. At prooemium illud in definitionem totius scientiae convertamus, quae eundem sensum reti-
neat et dicamus, scientia naturalis est, quae considerat prima principia rerum naturalium et simplicia corpora et mista tum in genere
tum species singulas, corpora tum animata tum inanima[9] usque ad
ultimas eorum species; tractatio verò eadem sequatur, quae nunc

order and make [it] definitive from compositive. Nothing more absurd than this, of course, can be imagined. For there is no one who does not know that this science is written using compositive order. To that, however, Aristotle adjoined a proem,[17] in which he did not, of course, expressly state everything to be considered in the whole science, but only [what was to be considered] in the *Nature Lectures* [i.e., the *Physics*]. To indicate their relationship to the other parts of that science was enough, for every proem is altogether discretionary, and the author can either utterly omit it or set out in it only the principal things, putting the rest aside, as Aristotle there set out the treatment to be made on first beginning-principles, without knowledge of which the other things cannot be well known.

Now we might fancy that Aristotle had there made a fuller 17 proem and had set out in it, in the following way, everything being considered in the whole of natural science: "In writing on natural science, we will speak in the first place about first beginning-principles, since without knowledge of them other things cannot be known. We will then discuss all natural things in order: first, of course, simples, next mixed [bodies], and these first generally and afterward specifically, until we come to the lowest species." Now once this proem of that science, in which appears everything to be treated in the whole science, has been established, if the whole subsequent treatment is maintained intact and unchanged and in the same sequence in which Aristotle wrote it, using what order will it be said to be written? Will it not be compositive? Without doubt! But were we to convert this proem into a definition of the science as a whole, [a definition] that would retain the same sense, and we were to say, "Natural science is what considers the first beginning-principles of natural things, and bodies simple and mixed, both in genus and in each of the species, bodies both animate and inanimate, all the way to their ultimate species," the same treatment as that which now exists would follow.

extat, dicat, quaeso, Galenus et illi, qui eum secuntur, quonam ordine tractatio illa scripta esse dicetur? Si definitivo, quid potest enunciari absurdius? Eadem enim tractatio est, eadem serie disposita, quae antè dicebatur scripta ordine compositivo, ne syllaba quidem in ea est mutata vel transposita, quomodo ergo potest in ea alius ordo esse genitus, si idem, qui prius erat, ordo servatur? Si verò dicant esse adhuc ordinem compositivum, ut revera esset, ipsi proprium suum dogma evertunt de ordine definitivo, nam ordini scientiae naturalis aptabitur definitio ab eis tradita ordinis definitivi et conditiones eius omnes, nec tamen erit definitivus.

18   Erit quidem primo loco proposita totius scientiae definitio sumpta à rebus omnibus in ea considerandis, quae postea ab authore in partes omnes resolvitur, cùm agat primùm de primis principiis, postea de simplicibus corporibus, deinde de mistis, et denique de aliis omnibus ordinatim, prout in ea definitione fuerant disposita. Cur ergo non erit ordo definitivus? Fateantur vel inviti ordinem hunc commentum esse et nulla ratione in speciebus ordinis numerandum.

19   Ex his colligimus in errorem manifestum lapsos esse nonnullos, qui inani argumento decepti putaverunt Avicennam, quòd in artis medicae traditione ab eius definitione auspicatus est, eam artem scripsisse ordine definitivo.

20   Quae sententia neque vera est neque verisimilis, siquidem ea quoque apparente° defensione destituti sunt, ad quam huius ordinis sectatores, ut mox considerabimus, confugere videntur, cùm dicunt proprium ordinis definitivi esse breviloquium, quale servat Galenus in Arte medicinali. Sed Avicenna cum magna sermonis prolixitate artem illam scripsit; ideò cur definitivo ordine tradita dicatur equidem nullam neque veram neque apparentem° rationem video; quòd enim à definitione inceperit, id nihil est, quoniam

Let Galen and those who follow him say, I pray: "In what order will this treatment be said to be written?" If [they say] in definitive, what more absurd could be proposed? For it is the same treatment, disposed in the same sequence, that was earlier said to be written using compositive order. Not even a syllable in it has been changed or transposed. So how can another order be engendered in it, if the same order as was there before is maintained? But if they say it is still compositive order, as in truth it would be, they overturn their own doctrine about definitive order. For the definition of definitive order conveyed by them and all its characteristics will be applicable to the order of natural science, and nevertheless it will not be definitive.

Of course, the definition of the whole science set out in the first 18 place will be taken from everything being considered in it [i.e., the science], which is afterward resolved by the author into all the parts, since he first deals with first beginning-principles, and afterward with simple bodies, and then with mixed, and lastly with everything else in order, as they were disposed in the definition. Why, therefore, will this not be definitive order? Let them acknowledge, even if reluctantly, that this order is a fiction and in no way to be counted among the species of order.

From all this we gather that some, deceived by [this] inane argument, lapsed into manifest error. They believed that Avicenna, 19 because in conveying the medical art[18] he commenced with its definition, wrote the art using definitive order.

This position is neither true nor likely, since indeed they also 20 lack even the specious° defense in which sectarians of this order, as we will soon consider, appear to take refuge, when they say that brevity is proper to definitive order, such as Galen maintains in *The Art of Medicine*. But Avicenna wrote his art with a great prolixity of discourse, and so I see no true or specious° reason why indeed it may be said to be conveyed using definitive order. The fact that he began with a definition is nothing, since all

omnes disciplinae, ut praediximus, à definitionibus exordium su-
munt, nec ob id ordine definitivo omnes scriptae sunt. Quòd autem non à quacumque definitione, sed à definitione artis medicae
inceperit Avicenna, id ordinem reddere definitivum non potest.
Dicant enim (quaeso) an ablata illa medicinae definitione[10] ars illa
Avicennae integra et perfecta maneat, ita ut legi et intelligi possit,
necne. Hoc quidem si inficientur, non est quòd adversus eos, qui
res manifestas negant, disputare velimus. Iam enim scientiae naturalis exemplo ostendimus cuiusque disciplinae definitionem aequè
posse apponi et auferri integra manente ipsa disciplina. Hoc autem
si verum est, ut certè est verissimum, auferamus illam Avicennae
definitionem, ars igitur integra manens quonam ordine scripta dicetur? Si dicant definitivo, ridiculi sunt, nam si absque definitione
datur ordo definitivus, etiam absque anima rationali dabitur homo.
Si autem alio, ergo neque cum illa definitione est ordo definitivus.

21   Horum autem omnium ratio ea est, quam paulò ante exponere
coeperamus,[11] quoniam eiusmodi definitio ipsius disciplinae est
instar prooemii, neque est tractationis pars ulla, proinde neque
ipsius disciplinae, propterea nullus ordo potest considerari inter
illam disciplinae definitionem et ipsam tractationem, sed totus
ordo in sola tractatione et in rerum consideratarum serie compre-
henditur, non in earum propositione, quam author in prooemio
facit. Est enim ordo respectus quidam conveniens partium cuiusque disciplinae inter se, partes illas intelligendo, quae res illius
disciplinae tractant et sunt verè partes, nimirum partes tractationis. Ordo enim totus penes harum seriem attenditur, non penes
prooemium, in quo nihil author tractat, sed solùm quae sunt
tractanda proponit. Idcircò ordine alicuius scientiae constituto

disciplines, as we said earlier, take their beginning from definitions. They are not on account of this all written using definitive
order. That Avicenna, moreover, began not with just any definition whatever but with the definition of medical art, cannot render
the order definitive. Let them say (I pray) whether, with this definition of medicine taken away, Avicenna's art would remain intact
and perfect, such that it could still be read and understood, or not.
Of course, if they deny this — there is nothing that we would want
to debate with those who deny manifest things. For we have now
shown by the example of natural science that the definition of any
discipline can equally be adjoined and removed, with the discipline
itself remaining intact. But now if this is true, and it certainly is
very true, let us remove that definition of Avicenna's: in what order will the art then be said to be written, assuming it remains
unchanged? If they say in definitive, they are ridiculous, for if
there is definitive order without a definition, so also will there be
man without a rational soul. If however [they say] in another, then
the order is not definitive with that definition either.

The reason for all these things is that which we began to lay out 21
a little earlier. Since a definition of this type, [a definition] of the
discipline itself, is like a proem but is not any part of the treatment and accordingly not [any part] of the discipline itself, there
can be no order to consider between that definition of the discipline and the treatment itself. The whole order, rather, is comprehended in the treatment alone and in the sequence of things being considered, not in the setting out of them that the author
puts in the proem. For order is some sort of appropriate relationship of the parts of any discipline to each other, understanding
by these parts, those that treat the things of the discipline and
truly are parts, that is, parts of the treatment. For the whole order
is observed in their sequence, not in the proem in which the author treats nothing, but only sets out what things are to be
treated. And accordingly, with the order of any science established,

nulla est eius scientiae definitio, quae toti scientiae anteposita possit[12] ordinem variare, ea enim non est pars necessaria ipsius disciplinae, sed extraneum quoddam et arbitrarium, à quo seu à cuius
relatione ad tractationem nullus potest ordo desumi. Sed solùm à
relatione et respectu partium tractationis ad se invicem.

22   Propterea etiam liber Galeni de Arte medicinali aliquo ordine
praeter definitivum scriptus est vel nullum penitus ordinem habet,
quia definitio illa medicinae non est tractationis pars, sed prooemii
locum habet, nam primum eius libri caput prooemium quidem
est, sed nimis commune et logicum potius, quàm medicinale et
cuilibet alii disciplinae aptari posset; definitio autem medicinae,
quae ponitur in principio secundi capitis, proprium est eius artis
prooemium, quod, quemadmodum diximus, nullum ordinem potest constituere. Quis autem ille ordo sit, quo Galenus in ea tractatione usus est, posterius dicemus, cùm enim definitivus ordo non
detur, is vel compositivus est, vel resolutivus, vel nullus, si praeter
hos duos alius ordo non reperitur.

: V :

In quo defensiones quaedam adversariorum reiiciuntur.

1   Per haec satis demonstratam esse puto definitivi ordinis vanitatem,
sed ut adversariis nullus, ad quem confugiant, locus relinquatur,
tollenda est quaedam responsio, qua ipsi uti videntur. Cùm enim
nos dicamus, non dari ordinem definitivum, siquidem in omni facultate, quae hoc ordine tradita esse dicatur, aliquis alius ordo
inspici potest; ipsi ut definitivum ordinem confirment et ita ab
aliis ordinibus separent, ut cum illis non appareat esse commistus,
dicunt ordinis definitivi proprium esse breviloquium, et propriam

there is no definition of the science that, placed before the whole science, could alter the order. For it [i.e., the definition] is not a necessary part of the discipline itself but is something external and discretionary; no order can be drawn from it or from its relation to the treatment. [This can be done] only from the relation and relationship of the treatment's parts to each other.

And so Galen's book *The Art of Medicine* is written using some 22 order other than definitive or has no order at all, since that definition of medicine is not part of the treatment, but has a place in the proem. For the first chapter of the book, of course, is a proem, but an overly general one, and it could be applied more to logic than to medicine and any other discipline. The definition of medicine, however, which is put in the beginning of the second chapter, is the proem proper to this art; as we said, this [definition] can establish no order. Now which order it is that Galen used in that treatment, we will say later. Since there is no definitive order, it is either compositive or resolutive or none, if no other order beyond these two is found.

: V :

In which some defenses of our opponents are rejected.

I hold that the emptiness of definitive order has been sufficiently 1 demonstrated by means of all this. But so that no place remains in which our opponents can take refuge, a sort of response that they appear to use has to be gotten rid of. For while we say there is no definitive order, since in every branch of learning that is said to be conveyed using this order some other order can be observed, they, to confirm definitive order and thus separate it from other orders so that it does not appear combined with them, say that brevity is

eius utilitatem esse ut ad memoriam conferat, quo fit ut liber Galeni de Arte medicinali propter eius brevitatem non dicatur alio ordine esse dispositus, quàm definitivo. Scientia verò naturalis, quam tradidit Aristoteles, dicatur ordinem habere compositivum, non definitivum, etiam si talem ei apponamus definitionem, qualem antea diximus, nam propter librorum multitudinem et totius scientiae prolixitatem non posset ille ordo dici definitivus, etiam si à tali definitione exordiretur. Scientia verò naturalis, quae nuper edita est ordine definitivo, non dicitur tradita ordine compositivo, sed potius definitivo propter eius brevitatem, quae ad memoriam maximè confert, quod quidem de illa, quam Aristoteles composuit, dicere minimè possumus.

2   Verùm hi, qui haec dicunt, si ea diligenter considerarent, ab hac puerili defensione desisterent, nos enim non inficiamur, breviloquio maximè iuvari memoriam, sed quòd brevitas et prolixitas sint essentiales differentiae, quae diversas ordinis species constituere aptae sint,[13] nemo eruditus vir deberet enunciare, siquidem id à ratione et ab ipsa ordinis natura est prorsus alienum. Est enim ab his petendum, an huius ordinis definitivi natura in breviloquio sit constituta an in definitione disciplinae tali, qualem suprà declaravimus. Si in breviloquio, ergò etiam sine illa definitione erit ordo definitivus, quod absurdissimum est, nam unde vocabitur definitivus, si à nulla definitione auspicabitur? Praeterea sequeretur quòd quocumque ordine, imò et quacumque inordinatione de aliqua re breviter tractantes diceremur ordinem servare definitivum, quod quidem nemo, ne puer quidem, assereret. Si verò in definitione, quod[14] et rationi et appellationi magis consentaneum est, ergò etiam sine breviloquio poterit esse ordo definitivus.

proper to definitive order and that its proper utility is that it contributes to memory. And hence it happens that Galen's book *The Art of Medicine*, on account of its brevity, is not said to be disposed using any order other than definitive. But the natural science that Aristotle conveyed is said to have compositive order, not definitive, even if we adjoin to it a definition such as we said earlier. For on account of the multitude of books and the prolixity of the whole science, this order could not be called definitive, even if it began with such a definition. And a natural science brought out recently using definitive order is said to be conveyed not using compositive order but rather using definitive, on account of its brevity, which contributes so well to memory. Of course, we cannot say this at all about the one [i.e. the science] that Aristotle composed.

Surely those who say these things, if they were to consider 2
them carefully, would part from this puerile defense. For we do not deny that memory is very well helped by brevity, but no learned man ought to propose that brevity and prolixity are the essential differentiae that are able establish different species of order, since this is utterly alien to reason and to the nature of order itself. For it has to be asked of them whether the nature of this definitive order is established by brevity or by such a definition of the discipline as [the one] we made clear above. If by brevity, then there will still be definitive order without that definition, which is most absurd. For why will it be called definitive, if it commences with no definition? Moreover, it would follow that in treating something briefly, using whatever order or even using no order, we would be said to maintain definitive order; no one, of course, not even a child, would assert this. But if instead [the nature of the definitive order is established] by definition, which agrees more with both reason and the appellation, then there could be definitive order even without brevity.

3   At certè neque in definitione neque in brevitate neque in prolixitate orationis ipsa ordinis natura consistit, sed in rerum consideratarum serie, earumque ad se invicem respectu et relatione.
Hinc hauriendae sunt differentiae, quibus ordo dividatur et diver-
sae eius species constituantur. Nam si propositam habeamus eandem scientiam easdem res et eadem serie dispositas tractantem, à
duobus authoribus scriptam, sed ab altero breviter, ab altero verò
prolixè, non est dicendum eam alio et alio ordine ab iis duobus
traditam esse. Sed idem est ordo ab utroque authore servatus;
quemadmodum dicere debemus de naturali philosophia ab Aristotele prolixè tradita et de eadem his temporibus scripta, ut multis
videtur, ordine definitivo. In hac enim idem ordo servatur, quem
servavit Aristoteles, quare compositivus est, non definitivus.

4   Cùm autem omnis ordo quatenus ordo est disciplinam faciliorem reddat et ad memoriam conferat, contrà verò inordinatio
faciat ut res difficulter memoriae mandentur. Magis tamen memoria retinemus ea, quae breviter dicta sint, quàm quae cum sermonis prolixitate, quia facilius est pauca recordari, quàm multa. Sed
in hoc non est constituenda alicuius ordinis natura, accidens enim
est, quod aequè cuilibet ordini competere potest, quare nulla ordinis species per hoc ab aliis ordinibus distinguenda est.

5   Quòd si quis dicat proprium esse ordinis definitivi ut non
solùm ipsius disciplinae definitio ante omnia proponatur, sed res
quoque omnes, quae in ea tractantur, per proprias definitiones
explicentur; id similiter non rectè dicitur. Ad methodum enim
potius, quàm ad ordinem pertinere videtur. Praeterea hoc idem in

But certainly, the very nature of order consists neither in defini- 3
tion nor in brevity nor in prolixity of speech, but in the sequence
of the things under consideration and in their relation and relationship to each other. It is from these that the differentiae, by
which order is divided and by which its different species are established, have to be drawn up. For if we have the same science set
out, treating the same things disposed in the same sequence, written by two authors, but by the one briefly and by the other prolixly, it must not be said that it is conveyed by these two, [one]
using one order and [the other] using another. Instead the same
order is maintained by each author, just as we ought to say about
the natural philosophy conveyed prolixly by Aristotle and about
the same written these days, as it appears to many, using definitive
order. For in this latter, the same order is maintained that Aristotle maintained; therefore, it is compositive, not definitive.

Now every order, insofar as it is order, renders the discipline 4
easier and makes it easier to remember; disorder, on the other
hand, does the opposite, so that things are committed to memory
with difficulty. Nevertheless we retain in memory more those
things that are said briefly than those [said] with a prolixity of
discourse, because it is easier to remember fewer things than
many. But the nature of some order is not to be established on this
basis, for it is an accident that could equally appertain to any order
whatever. Therefore no species of order is to be distinguished from
any other orders on account of this [brevity].

Now if anyone says that it is proper to definitive order, not only 5
that the discipline's definition is put forward before all else, but
also that everything that is treated in it is explicated by means of
proper definitions—this, similarly, is not correct to say. For it appears to pertain to method rather than to order. And we can note

philosophia naturali ab Aristotele tradita ordine compositivo animadvertere possumus, etenim nihil ferè in ea comperiemus, cuius
definitionem aut expressè aut saltem implicitè Aristoteles non ad-
ducat, omnium enim et substantiarum et accidentium perfecta
cognitio consistit in cognitione quid est, ut mox, cùm de methodis
loquemur, declarabimus.

6   Aliquos etiam audivi dicentes proprium esse ordinis definitivi
transire ab universalibus ad particularia, quoniam Galenus tradita
illa artis medicae definitione dividit statim salubria et insalubria in
corpora, signa et causas, quare haec sub illa definitione tanquàm
species sub gènere continentur.

7   Sed hanc esse non posse huius ordinis constitutricem differentiam ex iis, quae generaliter de ordine diximus, manifestum est,
ostendimus enim communem omnis ordinis conditionem esse ut à
communibus ad propria transitus fiat. Quare etiam in scientia naturali tradita ordine compositivo hunc ordinem conspicere possumus, nam licèt in libris Physicorum agat Aristoteles de primis
rerum naturalium principiis, tamen subiectum in iis libris est corpus naturale latissimè sumptum, quod postea ab Aristotele in
principio primi libri de Coelo dividitur in simplex et mistum; et
omnia aliorum librorum naturalium subiecta sub naturali corpore
tanquàm species sub genere comprehenduntur.

8   Sed nonne errorem, ac caecitatem suam detegunt, qui haec
dicunt, dum ordinis definitivi conditionem et naturam aliunde,
quàm à definitione desumunt? Ea enim certè in sola definitione, à
qua etiam appellationem hic ordo accepit, collocanda esse videtur.

9   Atqui propriamne[15] esse ordinis definitivi conditionem asserent
ut à definitione disciplinae non ex ipsis rebus considerandis, sed ex

the same thing in the natural philosophy conveyed by Aristotle using compositive order. Indeed we will find in it [i.e., his natural philosophy] almost nothing whose definition Aristotle does not adduce either explicitly or at least implicitly. For perfect knowledge both of all substances and of all accidents consists in knowledge of what something is (*quid est*), as we will soon make clear when we speak about methods.

Now I heard others saying that to pass from universals to par- 6 ticulars is proper to definitive order given that Galen, after convey- ing that definition of the medical art, at once divides healthy and unhealthy in bodies, signs, and causes; therefore, these are con- tained under that definition as species under a genus.

But it is manifest, from the things that we said generally about 7 order, that this cannot be the differentia constitutive of this order. For we showed that a common characteristic of every order is that a passage be made from common to proper things. And accord- ingly in natural science conveyed using compositive order, we can see this order. Now, granted that in the *Physics*, Aristotle deals with the first beginning-principles of natural things, nevertheless, the subject in these books is natural body taken in the broadest sense, which is afterward, in the beginning of the first book of *On the Heavens*,[19] divided by Aristotle into simple and mixed, and all the subjects of the other books on nature are comprehended under natural body, as species under a genus.

But do those who say these things not expose their error and 8 blindness when they draw what is characteristic of and the nature of definitive order from something other than definition? For it appears that certainly these things [i.e., the characteristic and na- ture of definitive order] have to be included in the definition alone, from which this order even gets [its] appellation.

Yet will they not assert that a characteristic proper to definitive 9 order is that the beginning is taken from the discipline's definition, which is conveyed not from the very things being considered but

earum genere tradita exordium sumatur, qualem eam Galeni definitionem esse diximus? Hoc autem, si benè expendatur, puerile est. Dicant enim, quaeso, si in ea definitione Galenus corpora et signa et causas expressisset, dicens, medicina est scientia salubrium, insalubrium et neutrorum, tum corporum tum signorum et causarum, fuissetne amplius ordo eius libri definitivus? Id certè negare vanissimum est et omni ratione carens, cur enim in definitione artis ipsum potius commune rerum tractandarum genus, quàm species ipsas nominare convenit? Et cur ab illa potius definitione per genus tradita, quàm ab ea, quae per species tradatur, constituitur ordo definitivus? Igitur si aequè hoc modo atque illo esset ordo definitivus, nullus relinquitur processus ab universalibus ad particularia, quem eius ordinis proprium esse comminiscuntur.

10 Dicunt etiam aliqui proprium esse ordinis definitivi incipere à principiis cognitionis, ut per hoc distinguatur à compositivo, qui est à principiis rei. Nec vident, hanc quoque esse communem omnium ordinum conditionem, imò in hac una totam ordinis naturam, quemadmodum ostendimus, esse constitutam. Ordo enim ab iis auspicatur, quorum cognitio ad aliorum cognitionem confert, hoc autem est à cognitionis principiis auspicari; idque in omni ordine videre possumus, nam ordo compositivus est à principiis rei quatenus sunt etiam principia cognitionis, ut asserit Aristoteles in principio primi libri Physicorum et praeter principia rei habet etiam quaelibet scientia propria cognitionis principia. In ordine autem resolutivo exordium sumitur à fine non ob aliam rationem, quam quia finis est principium totius cognitionis, non enim est principium rei. Nulla igitur inter ordinem definitivum et alios duos differentia manet, nisi aliqua propria conditio afferatur, quae declaret quaenam sint haec principia cognitionis, à quibus

from their genus, as we said Galen's definition is? But this, if it is well weighed, is puerile. For let them say, I pray, if Galen had expressly stated "bodies," "signs," and "causes" in the definition, saying, "Medicine is the science of the healthy, of the unhealthy, and of neither, both of bodies and of signs and causes," would the order of the book no longer have been definitive? To deny this is certainly most vain and lacking all reason. For why in the definition of an art is it appropriate to name the common genus itself of the things being treated rather than the species themselves? And why is definitive order established by the definition conveyed by means of the genus rather than by that which is conveyed by means of the species? If, therefore, the order would be equally definitive in this way and in that, then no proceeding from universals to particulars — what they allege to be proper to this order — remains.

Now some also say that it is proper to definitive order to start   10
from the beginning-principles of knowledge, and that by means of this it is distinguished from compositive, which is from the beginning-principles of the thing. But they do not see that this characteristic is also common to all orders. Indeed in this one [characteristic] the whole nature of order is constituted, just as we showed. For order commences with things knowledge of which contributes to knowledge of other things. This, however, is to commence with the beginning-principles of knowledge, and this we can see in every order. For compositive order is from the beginning-principles of something insofar as they are also beginning-principles of knowledge, as Aristotle asserts in the beginning of the first book of the *Physics*,[20] and beyond the beginning-principles of the thing any science also has its own proper beginning-principles of knowledge. But now in resolutive order, the beginning is taken from the end, not for any other reason than that the end is the beginning-principle of knowledge as a whole, but it is not the beginning-principle of the thing. No difference,

auspicari proprium sit ordinis definitivi. Haec autem quaenam sit non video, nisi ipsam disciplinae definitionem dicant, à qua[16] ordo vocatur definitivus, nam revera in hac una persistere deberent et ab ea sola propriam huius ordinis naturam desumere.

11   Quid igitur? Dicentne ordinis definitivi naturam in hoc esse constitutam, ut incipiat à tali principio cognitionis, non ab alio, scilicet ab ipsa disciplinae definitione?

12   Sed haec quomodo cognitionis principium dicatur non apparet, nisi dicamus quatenus admonet nos quaenam sint res in ea disciplina considerandae. Sed hoc cognitionis principium appellare ridiculum est, ex eo enim in nullius rei ignotae cognitionem ducimur. Idque etiam prooemium omne, in quo author dociles auditores reddat, praestare manifestum est, nec propterea dicimus disciplinam omnem tale prooemium habentem[17] scriptam esse ordine definitivo.

13   Revera enim neque prooemium neque disciplinae definitio potest cognitionis principium appellari, cùm tractationis pars minimè sit et arbitratu nostro deleri atque amoveri possit integra manente ipsa disciplina. Sed principia cognitionis illa esse dicuntur, ex quorum cognitione aliorum omnium cognitio pendet et quibus sublatis totam illam disciplinam corruere necesse est.

therefore, remains between definitive order and the other two, unless some proper characteristic is brought forward that makes clear what these beginning-principles of knowledge are, with which to commence is proper to definitive order. I do not see what this could be, unless they say it is the discipline's definition itself, by which the order is called definitive, for, in truth, they ought to stick with this one thing and draw from it alone the nature proper to this order.

So what? Will they not say that the nature of definitive order is 11 constituted in this, that it starts from such a beginning-principle of knowledge, not from another, that is, from the definition itself of the discipline?

But it is not apparent in what way this is called a beginning- 12 principle of knowledge, unless we say, insofar as it points to what sort of things have to be considered in the discipline. But it is ridiculous to call this a beginning-principle of knowledge, for from it we are led to knowledge of nothing unknown. And it is manifest that every proem in which an author renders his listeners ready to learn does this. But we do not say that, on that account, every discipline having such a proem is written using definitive order.

For in truth neither the proem nor the definition of a discipline 13 can be called a beginning-principle of knowledge, since it is not part of the treatment at all and can be deleted and removed at our discretion, the discipline itself remaining intact. But beginning-principles are said to be those things on the knowledge of which the knowledge of all others depends and without which the whole discipline necessarily falls apart.

: VI :

In quo propria sententia de divisione,
ac de speciebus ordinis exponitur.

1 Hactenus adversus alios de divisione ordinis ac de ordinum numero satis superque disputavimus. Nunc nitendum nobis est veras
invenire differentias, quibus ordinem in species dividamus, neque
dubitandum° nostram hac in re sententiam adducere, etsi novam
et post Aristotelem nemini, ut videtur, ad haec usque tempora
cognitam.

2 Diximus autem[18] finem omnis ordinis esse nostram meliorem,
ac faciliorem cognitionem, in hoc tota ordinis natura et essentia
posita est, ab hoc igitur sunt sumendae differentiae ordinem divi-
dentes, per quas ordinum numerus declaretur.

3 Quoniam enim ordine utimur non ob aliam causam, quàm ut
cognoscamus, scopus hic in duos dividitur, aut enim cognitionem
propter seipsam quaerimus et eam adepti quiescimus nec aliquid
praeterea quaerimus, quod evenit in solis scientiis contemplativis.
Aut ipsam non propter se quaerimus, sed ut acquisitam ad operationem dirigamus, quod in aliis omnibus disciplinis contingit,
quae contemplativae non sunt, nempè in morali philosophia et in
artibus omnibus cunctisque aliis facultatibus, quae excogitari possint, in his enim cognitio quidem quaeritur, sed operatio, cuius
gratia cognitio quaeritur, ultimus est ac praecipuus finis. Operationem autem intelligo earum rerum, quae ab homine per liberum
suae voluntatis arbitrium gigni ac fieri possunt, quemadmodum
fusè declaravimus aliàs in libro nostro de Natura logicae.

: VI :

In which the proper position regarding the division and species of order is laid out.

Up until now we have debated with others more than enough   1
about the division of order and the number of orders. We now
have to endeavor to discover the true differentiae by which we divide order into species. And we should not hesitate° to adduce our
position in this issue, although it is new and—all the way to our
own times it appears—known to no one after Aristotle.

Now we said the end of every order is our better and easier   2
knowledge. In this [end] is located the whole essence and nature
of order. From this [end], therefore, the differentiae dividing order
have to be taken, and by means of them the number of orders may
be made clear.

Now since we use order for no cause other than that we may   3
know, this goal is divided into two. For either we inquire after
knowledge for its very own sake, and then, having obtained it, we
rest and inquire after nothing else; this happens only in contemplative sciences. Or we inquire after it not for its own sake, but so
that we may direct toward practical activity what is acquired. This
happens in all other disciplines, those that are not contemplative,
that is, in moral philosophy, and in all the arts, and in all other
branches of learning that could be imagined. In these, of course,
knowledge is inquired after, but practical activity, for the sake of
which the knowledge is inquired after, is the ultimate and principal end. I understand practical activity to be that regarding the
things that can be engendered and made by a person by means of
the unconstrained choice of his will, as we made clear at length
elsewhere, in our book *On the Nature of Logic*.[21]

4   Quoniam igitur aliae disciplinae solam rerum tractandarum cognitionem pro fine habent, in aliis verò finis aliquis proponitur à nobis producendus, qui à rebus tractandis et ab earum cognitione diversus est; ex his duabus disciplinarum naturis, duobusque earum diversis scopis duo diversi ordines oriuntur, nam si ipsam propter se rerum quaerimus, oritur necessitas ordiendi à primis principiis. Si verò non propter se, sed propter operationem et alicuius finis adeptionem, cogimur ab illius finis notione auspicari et ad prima principia investiganda procedere, à quibus operationem postea incepturi sumus, duo igitur soli ordines dantur, qui ex ipsa rerum cognoscendarum natura, prout à nobis cognoscendae sunt, deducuntur,[19] unus est compositivus, qui à primis principiis inchoando progreditur ad posteriora principia° et ad ea, quae ex principiis constant et à simplicibus ad composita, ut perfecta rerum cognitio tradatur. Alter est resolutivus, qui proposito ultimo fine agendo vel efficiendo à nobis, progreditur ad prima principia indaganda, per quae finem illum postea producere et comparare possimus.[20] Propterea scientiae contemplativae alio ordine tradi nequeunt, quàm compositivo, ipsisque resolutivus ordo nulla ratione aptari potest. Artes verò ac disciplinae aliae omnes solo resolutivo uti possunt, minimè verò compositivo, qua in re omnes video fuisse deceptos, qui de hac re scripserunt, existimantes naturalem philosophiam posse alio ordine tradi, quàm compositivo; artem autem medicam non minùs compositivo, quàm resolutivo. Haec autem omnia ita facile[21] est demonstrare, ut mirandum profectò sit tot viros in philosophia atque in medica arte praeclaros

Since some disciplines, therefore, have for an end only knowl- 4
edge of the things being treated, but in others the end is something set out to be produced by us that is different from the things
to be treated and from knowledge of them, from these two natures
of disciplines and their two different goals, two different orders
arise. Now if we inquire after that [knowledge] of things for its
own sake, the necessity of ordering from first beginning-principles
arises. But if not for its own sake and instead for the sake of practical activity and the obtaining of some [other] end, we are forced
to commence with a notion of that end and to proceed to the first
beginning-principles being investigated, from which afterward we
will start the activity. There are, therefore, only two orders that are
deduced from the very nature of the things to be known, in that
they are to be known by us. One is compositive, which, in starting
with first beginning-principles, progresses to later principles° and
to those things that are composed out of beginning-principles, and
from simples to composites, so that perfect knowledge of things
may be conveyed. The other is resolutive, which, once the ultimate
end to be done or effected by us is set out, progresses to the first
beginning-principles being tracked down and by means of which
we can afterward produce and procure that end. And so contemplative sciences cannot be conveyed using any order other than
compositive; resolutive order cannot be applied to them in any
way. But arts and all other disciplines can use only resolutive, not
compositive at all. As I see it, all who wrote on this issue, judging
that natural philosophy can be conveyed using any order other
than compositive — and on the other hand that medical art [can be
conveyed] no less by compositive than by resolutive [order] — were
deceived. Now all this is so easy to demonstrate that it surely has
to be wondered how so many famous men in philosophy and in

hanc veritatem non cognovisse, quam nobis res ipsa declarat. Nos enim non aliunde, quàm ex ipsa rei natura argumentum sumpturi sumus ad ea omnia, quae diximus, comprobanda.

: VII :

In quo ostenditur scientias contemplativas non posse alio ordine tradi, quàm compositivo.

1 Quod scientiis contemplativis alius ordo non conveniat, quam compositivus ratio Aristotelis demonstrat, quam legimus in ipso statim initio primi libri Physicorum, quem locum quotidie philosophi tractant, nec tamen, quae sit vis verborum Aristotelis benè perpendunt. Ostendere ibi vult Aristoteles in tradenda scientia naturali esse auspicandum à primis rerum naturalium principiis, quod quidem nihil aliud est, quàm eam scientiam esse disponendam ordine compositivo, ad hoc autem demonstrandum sumit argumentum à fine, finis enim cuiusque scientiae speculativae est sola cognitio, ac perfecta rerum scientia, haec autem haberi non potest nisi cognitis primis principiis, id est primis in eo genere, ex quibus caetera omnia eiusdem generis conflantur et constant. Res igitur naturales cognosci non poterunt nisi cognitis primis ipsarum principiis, quae elementa propriè dicuntur, ab horum itaque tractatione exordiendum est. Ea enim in disciplinis anteponenda sunt, quae ad aliorum cognitionem adipiscendam vel necessaria vel

the medical art have not known this truth, [a truth] made clear to us by the thing itself. For to confirm everything we have said, we will take [our] argument from nothing other than the very nature of the thing.

: VII :

In which it is shown that contemplative sciences cannot be conveyed using any order other than compositive.

Aristotle's reasoning demonstrates that no order other than compositive is appropriate for contemplative sciences. We read this at once at the very start of the first book of the *Physics*,[22] in a passage philosophers treat every day. But they do not well examine what the force of Aristotle's words is. Aristotle wants to show there that in conveying natural science, commencing has to be from the first beginning-principles of natural things, which, of course, is nothing other than that the science has to be disposed using compositive order. And to demonstrate this he takes [his] argument from the end. For the end of any speculative science is just knowledge, and perfect scientific knowledge, of things. This, however, cannot be had except from known first beginning-principles (first, that is, in the genus), out of which everything else in the same genus is assembled and composed. Natural things, therefore, cannot become known except from their known first beginning-principles, what are properly said to be elements. And so one has to begin with a treatment of them. For in disciplines, those things have to be placed first that are necessary or at least useful for obtaining

1

saltem utilia sunt. At primorum principiorum cognitio est necessaria pro rerum naturalium perfecta cognitione, ergo cogimur
primo loco de principiis primis agere, qui est ordo compositivus.

2   Qui igitur dicunt, scientiam naturalem posse tradi etiam ordine
resolutivo, si authoritati Aristotelis non acquiescunt, ad ipsius argumentum, si possunt, respondeant.

3   Nam si res naturales alio ordine cognosci possent, quàm compositivo, ratio illa Aristotelis vana esset et maiorem propositionem
falsam haberet, ea tamen non modò vera est, sed axioma notissimum, est enim ipsa scientiae definitio ab Aristotele tradita in
principio primi libri Posteriorum et omnium hominum consensu
comprobata, quae ut in scientia naturali, ita in aliis omnibus contemplativis scientiis locum habet, in omnibus enim eadem ratio
viget. Ideò non benè maiorem illam intellexere illi, qui existima-
runt dictionem illam 'methodos' in ea propositione positam latissimè patere et ad omnes penitus disciplinas extendi. Hoc enim si
admitteretur, sequeretur Aristotelem ea ratione probare omnes
disciplinas scribendas esse solo ordine compositivo.

4   Nam in arte aedificatoria dantur aliqua prima principia et elementa, è quibus alia conflantur et constant, ergo si ex eorum cognitione cognitio aliorum penderet, sequeretur in ea arte tradenda à
lapidibus auspicandum esse.

5   Aristoteles tamen contrarium asserit in contextu 23 septimi libri
Metaphysicorum, ubi dicit in arte aedificatoria tradenda à domo,
quae finis est, incipiendum esse, et ita ad principia progrediendum,
non à principiis. Manifestum est igitur maiorem illam non esse
intelligendam nisi de scientiis contemplativis, quas solas significavit Aristoteles dicens, methodos, non quòd aliae non possint

knowledge of other things. But knowledge of first beginning-principles is necessary for perfect knowledge of natural things. Therefore we are forced to deal in the first place with first beginning-principles; this is compositive order.

Let those, therefore, who say that natural science can be conveyed using resolutive order also, if they do not acquiesce in Aristotle's authority, respond to his argument, if they can. 2

For if natural things could be known using some order other than compositive, then Aristotle's reasoning there would be vain and have a false major premise;²³ but in reality that is not only true, but a most well-known axiom. For it is the very definition of scientific knowledge conveyed by Aristotle in the beginning of the first book of the *Posterior* [*Analytics*]²⁴ and confirmed by the consensus of all men, and it has a place in all other contemplative sciences just as in natural science, for the same reasoning applies in all. And so those who judged that locution "method," placed in the [major] premise,²⁵ as spreading out very broadly and completely extending to all disciplines, did not understand that major [premise] very well. And if this should be admitted, it would follow that Aristotle proved by this reasoning that all disciplines should be written using compositive order alone. 3

For in the art of building, there are some first beginning-principles and elements out of which other things are assembled and composed. Therefore if knowledge of the others depended on knowledge of them, it would follow that, in conveying this art, one has to commence with stones. 4

Aristotle asserts the contrary, however, in text no. 23 of the seventh book of the *Metaphysics*,²⁶ where he says, in conveying the art of building one has to start from the house, which is the end, and thus progress to beginning-principles, not from beginning-principles. It is manifest, therefore, that that major [premise] is not to be understood as about anything except contemplative sciences. Aristotle, saying "method,"²⁷ indicated only these, not 5

vocari methodi, sed quia propositio illa restringitur ab illis verbis,
cognoscere et scire, tum quia propriè sumpta scientia locum non
habet nisi in contemplativis, ubi res aeternae et necessariae tractantur, cùm reliquae disciplinae versentur in contingentibus, quae
à nobis fieri et non fieri possunt; tum etiam quia in caeteris scopus
non est cognitio, sed operatio, dum igitur id, quod in eis praecipuum est, respicimus, illae cognitionem non quaerunt, sed hoc
speculativarum est proprium, quare modus ipse loquendi significat
Aristotelem ibi de illis tantùm methodis loqui, quae finem habent
praecipuum scientiam, ut sensus sit, in omnibus disciplinis, quarum scopus sit cognitio et scientia et quae habeant aliqua prima,
ex quibus alia constent, cognitio ipsa et scientia non acquiritur nisi
ex illorum primorum cognitione, quod quidem praedicatum de
aliis disciplinis ab Aristotele non pronunciatur, quoniam illae sub
subiecto illius propositionis non continentur.

6   Ratio igitur haec Aristotelis est efficacissima ad demonstrandum quòd scientiae speculativae solo ordine compositivo tradi
possint.

7   Idem confirmare possumus argumento sumpto ab ordine resolutivo, qui scientiis speculativis nullo pacto aptari potest, hic enim
à fine auspicatur, ut omnes concedere videntur, ab[22] omnibus enim
hic ordo vocatur à notione finis. Sed in scientiis speculativis quis-
nam finis praeter subiectarum rerum cognitionem proponitur?
Certè nullus, at ipsa rerum scientia non est finis talis, qui possit
ordinem resolutivum constituere, qua in re decepti sunt nonnulli,
scientia enim à rebus ipsis sciendis reipsa non differt, sed idem est,

because others could not be called methods, but because that premise is restricted by those words, "to know (*cognoscere*)" and "to know scientifically (*scire*)" — because scientific knowledge (*scientia*), taken properly, has no place except in contemplative [disciplines], where eternal and necessary things are treated, while the remaining disciplines are concerned with contingent things, which can be made or not made by us; and also because in the other [disciplines], the goal is not knowledge, but practical activity. When, therefore, we give regard to that which is principal in these, [we conclude that] they [i.e., the latter disciplines] do not inquire after knowledge; it is, however, proper to the speculative [disciplines]. And so the very mode of speaking indicates that Aristotle there speaks only about those methods that have scientific knowledge as the principal end. So the sense [in that passage] is that in all disciplines whose goal is knowledge (*cognitio*) and scientific knowledge (*scientia*), and that have some first things out of which other things are composed, the knowledge (*cognitio*) and scientific knowledge (*scientia*) are not acquired except from knowledge (*cognitio*) of those first things; this predicate, of course, is not said by Aristotle of other disciplines, since those are not contained under the subject of that premise.

This reasoning of Aristotle's, therefore, is most effectual for 6 demonstrating that speculative sciences can be conveyed using compositive order alone.

We can confirm the same thing by an argument taken from 7 resolutive order, which [order] can in no way be applied to speculative sciences. For it commences with the end, as everyone appears to concede; for by everyone, this order is designated on the basis of a notion of the end. But in speculative sciences, what end is set out other than knowledge of things of the subject? Certainly none. And the scientific knowledge itself of things is not such an end that it could establish resolutive order. In this issue some have been deceived. For scientific knowledge does not differ in reality

quod duobus modis sumitur, nempè vel in animo secundùm esse cognitum (ut ita loquamur) vel extra animum secundùm esse reale, quare nullus ordo potest considerari à scientia ipsa ad res; sed neque à desiderio nostro sciendi ad ipsius scientiae adeptionem seu constitutionem, ut aliqui finxerunt, qui speciem quandam ordinis resolutivi invenerunt à fine ad totam artem et à tota arte ad partes, eamque philosophiae naturali accommodare ausi sunt. Quam sententiam equidem figmento similem esse semper existimavi, quandoquidem id, quod dicunt cupiditatem cognoscendi res naturales, quas intuemur, nos movere ad scientiae naturalis constitutionem, non rectè dicitur, neque verum[23] est, nisi sano modo intelligatur, nam quando trahimur desiderio cognoscendi res naturales, tunc ipsarum cognitionem nondum habemus, quare optamus quidem earum scientiam[24] adipisci, sed artem constituere sive disciplinam, in qua earum cognitionem aliis tradamus, id certè non possumus desiderare antequàm nos ipsi ea cognitione potiamur. Sed ea potiti[25] trahi possumus desiderio iuvandi alios et talem facultatem scribendi vel docendi. Ubinam igitur locum habet hic nuper inventus ordo resolutivus?

8   Primùm quidem dum nos ipsi discimus laborando et contemplando, qua utamur resolutione non video, quòd enim à sciendi cupiditate ad res ipsas contemplandas moveamur, id nihil aliud est, quàm nos à rebus ipsis sciendis moveri ad earum contemplationem, ut eas sciamus, ipsae igitur sunt et finis et subiectam totius nostrae contemplationis, in quo versamur et in principio scientiae et in medio et in fine. Quare est progressio à fine ad finem, ab

from the things themselves so known. Rather, the same thing is taken in two ways, that is, either in our soul according to known being (*esse cognitum*) (as we say) or outside our soul according to real being (*esse reale*). Hence no order can be considered from scientific knowledge itself to the thing, nor from our desire for knowing scientifically to the obtaining and constituting of the scientific knowledge itself, as some fancied, who [thought they had] discovered a sort of species of resolutive order from the end to the whole art and from the whole art to the parts, and dared to apply it to natural philosophy. For my part, I always judged this position to be like a figment, since what they say, that a longing for knowing the natural things that we see moves us to the constitution of natural science, is not said correctly, nor is it true—unless it is understood in a sound way. For when we are driven by the desire for knowing natural things, we still do not yet have knowledge of them. And so, of course, we want to obtain scientific knowledge of them, but we certainly cannot desire to constitute the art or the discipline in which we convey knowledge of them to others before we ourselves get hold of that knowledge. But having gotten hold of it, we can be driven by desire to help others and to write or teach such a branch of learning. Where on earth, therefore, does this recently discovered resolutive order have a place?

First, of course, I do not see what resolution we are using, when we ourselves learn by laboring and contemplating. For that we are moved to contemplate the things themselves from a longing for knowing scientifically—this is nothing other than that we are moved by the things to be scientifically known to contemplation of them, so that we may scientifically know them. These things, therefore, are both the end and the subject of our whole contemplation, that with which we are concerned at the beginning and the middle and the end of the science. Therefore it is progression from the end to the end, from one thing to the selfsame thing,

eodem ad idem, quia[26] in omni nostrae contemplationis parte scientiam aliquam adipiscimur. Haec igitur non est resolutio, quia tunc resolvimus, quando à fine progredimur ad alia quaedam ab eo diversa et ipso priora, quod quidem hîc ne fingere quidem possumus.

9 Sed ipsa quoque appellatio hunc errorem declarat, ordinem enim resolutivum omnes vocant ordinem à notione finis, finem ergo ante omnia notum esse oportet, deinde progressum fieri ad alia praeter finem. Hîc autem nullum praecognoscimus finem, quia res ipsae sciendae sunt finis, has autem ignoramus et ignoratio est causa appetitus, quomodo igitur procedimus à notione finis? Quòd si dicant praecognosci nomen hoc, scientia, hoc certè nihil est, haec enim est praecognitio logicae, à logica autem ad philosophiam et ab intentionibus ad res procedere non est ordine uti resolutivo, neque omninò ordine aliquo, ordo enim est quidam respectus inter ea, quae sint eiusdem generis, ut à principiis rerum naturalium ad effectus ipsos naturalis; et à sanitate ad causas, per quas conservari sanitas queat vel amissa recuperari. Proinde ordo est à re tractanda ad aliam rem tractandam seu ab intentione ad intentionem, nam etiam in disciplinae logicae traditione aliquem servari ordinem dicimus. Sed ab intentione ad rem progredi non est ordo, de quo nobis tractatio proposita est. Non est enim ordo, quo disciplina ipsa disponitur, sed potius progressus quodammodo à prooemio ad tractationem ipsius disciplinae.

10 Postquàm autem nos ipsarum rerum cognitionem adepti sumus, ut rerum naturalium et desiderio movemur scribendi naturalem scientiam ad aliorum utilitatem, adhuc non video ordinem hunc resolutivum à fine ad totam scientiam et à tota scientia ad partes. Ille enim, qui iam rerum cognitionem in animo habet,

because we obtain some scientific knowledge in each part of our contemplation. This, therefore, is not resolution, because we resolve when we progress from the end to some other things different from and prior to it—which, of course, we cannot even fancy here.

But the appellation itself also makes the error clear. For everyone calls resolutive order the order based on a notion of the end. The end, therefore, must be known before all else. The progression then occurs to other things besides the end. But here we know no end beforehand, because the very things to be scientifically known are the end, but we are ignorant of these, and ignorance is the cause of [our] eagerness. In what way, therefore, do we proceed from a notion of the end? Now if they say that this name, "science," is known beforehand, certainly this is nothing. For this is prior knowledge of logic. And to proceed from logic to philosophy and from intentions[28] to things is not to use resolutive order or any other order whatsoever. For order is some sort of relationship between things that are of the same genus, as [for example] from beginning-principles of natural things to the natural effects, or from health to the causes by means of which health can be conserved or, if lost, restored. Accordingly, order is from something being treated to something else being treated, or from intention to intention, for even in conveying the discipline of logic, we say that some order is maintained. But to progress from an intention to a thing is not the order in which the treatment is set out before us. For it is not the order in which the discipline itself is disposed. Rather, it is in some way a progression from the proem to the treatment of the discipline itself.

Now after we have obtained knowledge of the things themselves, as [for example] of natural things, and we are moved by the desire to write natural science for the utility of others, still I do not see that this is resolutive order from an end to the whole science and from the whole science to the parts. For he who now has

9

10

quando vult eam cum aliis hominibus communicare vel per vocem vel per literas, statim considerat quo ordine res ipsas tractandas disponere debeat. Quòd autem à desiderio alienae utilitatis trahi ad scientiam scribendam sit ordine uti resolutivo, quemadmodum isti comminiscuntur, mihi quidem ridiculum esse videtur et confutatione indignissimum, siquidem ordo, de quo nobis in praesentia sermo est, logicum instrumentum est, quod ad res melius cognoscendas confert et quo eruditi ac literati viri utuntur, à fine autem ad aliquid agendum moveri non modò omnium hominum, sed brutorum quoque animalium est, imò et stirpium et rerum quoque omni anima carentium, lapis desiderio naturali inferi loci ductus deorsum fertur, quid igitur? Utiturne lapis ordine resolutivo? Quòd autem aliud sit à fine moveri, aliud sit[27] ordinem servare resolutivum, omnis operatio tum naturae tum artis ostendit.

11   Naturam enim in rerum productione ordine uti compositivo manifestum est, siquidem à simplicibus ad composita progreditur, tamen quicquid agit, agit propter finem. Aedificator quoque domum construens agit propter finem, fine enim aliquo moventur, ordine tamen compositivo utitur dum aedificat, nec aliquis unquàm dixit in ipsa naturae et artis operatione commistum esse cum ordine compositivo alterum ordinem resolutivum, quòd à fine moveantur.[28] Quia ordinem servare dicimur in iis, quae agimus vel producimus, at moveri à fine non est aliquid producere vel operari. Sed quando agere incipimus propter finem, tunc in nostra operatione ordinem aliquem servamus, dum hoc prius facimus, vel tractamus, postea verò illud. Ordo igitur resolutivus non servatur quando à fine movemur, sed quando de fine prius agimus, postea

knowledge of things in [his] soul, when he wants to communicate it to other men, either through word or through writing, at once considers in what order he ought to dispose the very things being treated. But that this — to be driven to write a science by a desire for others' utility — is to use resolutive order, as they allege, appears to me, of course, to be ridiculous and most unworthy of refutation, since order, which our discourse is at present about, is a logical instrument that contributes to knowing things better and is used by learned and lettered men; but to be moved by an end to do something is [characteristic] not only of all men but also of all brute animals and even of plants and also of things lacking any soul. A stone, led by natural desire for a lower place, is carried downward. So what? Does the stone use resolutive order? Every activity of both nature and art shows that to be moved by an end is one thing, to maintain resolutive order is another.

It is manifest that nature uses compositive order in the produc- 11 tion of things, since it progresses from simples to composites. Nevertheless, whatever it does, it does for the sake of an end. A builder constructing a house also acts for the sake of an end. They are [both] moved by some end. Nevertheless, he uses compositive order when he builds, and no one has ever said that in the very activity of nature and of art, since they are moved by an end, another order — resolutive — is combined with compositive order. Though we are said to maintain order in the things that we do or produce, to be moved by an end is not to produce or practice something. But when we start to act for the sake of some end, then we maintain some order in our activity, when we first make or treat this, and then afterward that. Resolutive order, therefore, is not maintained when we are moved by an end, but when we first deal with an end and then afterward with those things that are before the end. For to be moved by an end is not to deal with the end, and whoever is moved by an end can start a treatment from first beginning-principles and produce the end only in the

verò de iis, quae sunt ante finem, moveri enim à fine, non est de fine agere et potest ille, qui à fine movetur, tractationem incipere à primis principiis, ultimo autem loco finem producere, ut patet in arte et in natura, quando agunt et producunt aliquid.

12   Vanus est igitur ille ordo resolutivus à fine ad totam artem et à tota arte ad partes, neque enim à tota arte ad partes aliquis unquàm procedit, cùm enim totum non sit nisi partium collectio, ars tota nihil est quando partes adhuc non extant, quare nullus alius fit progressus, quàm à parte artis ad partem, partibus autem omnibus iam compositis, tunc tota ars extare dicitur.

13   Sed ut admittamus etiam totam artem ante suarum partium traditionem modo quodam in animo authoris scientis existere, tamen ab ea ad partes progredi non est resolvere, sed dividere potius et disponere, scientiam enim naturalem traditurus cogitat in primis in quot partes tractationem illam dividere debeat et quo ordine partes illae sint disponendae. Talem authoris praemeditationem an liceat ordinem appellare resolutivum à tota arte ad partes, quisque rationis compos considerare potest, nam mihi supervacaneum esse videtur hac in re amplius morari. Satis igitur constat, in scientiis contemplativis nullum alium finem proponi, quàm scientiam rerum tractandarum et hunc non posse ordinem resolutivum constituere, cognitio enim ubi sola scopus ac finis est, cogit ordinem servare compositivum, ut ratio illa Aristotelis demonstrat, quam in ipso primi Physicorum initio legimus et cuius vim paulò antè perpendimus.

14   Restat ut aliquis dicat posse scientiam speculativam disponi ordine resolutivo, si ordo servetur compositivo contrarius, ut si scientiam naturalem auspicemur ab ultimis effectis et ab his ad proxima principia progrediamur, deinde ad remota, et tandem in remotissimis ac primis desinamus, à quibus incipit ordo compositivus, ut si ordiamur à tractatione ultimarum specierum, veluti hominis,

last place, as is patent in art and in nature, when they deal with and produce something.

This resolutive order, therefore, from an end to the total art and 12 from a whole art to the parts, is vain, for never does anyone proceed from a whole art to [its] parts. For since a whole is nothing but a gathering of parts, the whole art is nothing when the parts do not yet exist. Therefore no progression occurs other than from a part of the art to a part. When all the parts have been composed, then the whole art is said to exist.

But while we even admit that in some way the whole art exists 13 before the conveying of its parts in the soul of the author who knows [them] scientifically, nevertheless, to progress from it to the parts is not to resolve, but rather to divide and to dispose. For someone about to convey natural science ponders in the first place into how many parts he ought to divide that treatment and in what order those parts have to be disposed. Whether such premeditation by the author may be called resolutive order from the whole art to the parts, anyone in possession of reason can consider, but it appears to me superfluous to tarry further on this issue. This, therefore, continues to be sufficient: In contemplative sciences no other end is set out than scientific knowledge of the things being treated, and this cannot constitute resolutive order. For knowledge, where it is the only goal and end, forces [us] to maintain compositive order, as that reasoning of Aristotle's that we read in the very start of the first [book] of the *Physics* and whose force we examined a little earlier demonstrates.

It remains that someone could say that speculative science can 14 be disposed using resolutive order, if the order contrary to compositive were maintained; as if we were to commence natural science with the ultimate effects, and from these progress to the proximate beginning-principles and then to remote [ones], and finally to finish in the most remote and first [ones], from which compositive order starts; as if we were to begin with a treatment of

bovis, asini et aliarum omnium, mox transeamus ad partes hete-
rogeneas, deinde ad homogeneas, ab his ad tractationem de quatuor elementis et tandem perveniamus ad primae materiae et formae considerationem, à qua auspicatur scientia naturalis ab
Aristotele tradita ordine compositivo.

15   Sed qui tali ordine resolutivo putant naturalem scientiam tradi
posse, ignorare videntur quid sit ordo resolutivus, hic enim (ut ipsi
quoque confitentur) est à notione finis, at infimae species possunt
quidem vocari ultimum, sed non finis, quia non omne ultimum est
finis, ut docet Aristoteles in secundo libro Physicorum. Res autem
haec ut intelligatur sciendum° est totam naturalem philosophiam
ab Aristotele traditam eum ordinem servare, qui est ab universis
ad singula, quemadmodum ipse protestatur in prooemio primi libri Physicorum. Primùm enim in octo libris Physicorum considerat corpus naturale in sua maxima amplitudine acceptum, eiusque
principia et affectiones declarat, postea in libris de Coelo sumit
considerandum corpus simplex, quae est prima species corporis
naturalis: deinde corpus mistum latè sumptum, mox eius species
ordinatim usque ad ultimas. Propterea multi ab eo, quod primo
aspectu apparet, decepti crediderunt ultimarum specierum cognitionem esse ultimum finem philosophi naturalis, cùm generum
cognitio propter specierum cognitionem quaeri videatur.

16   Verùm non ita se res habet, ut enim admittamus ipsas species
esse talem finem, qui possit[29] ordinem resolutivum constituere,
quod quidem minimè verum est, tamen dicimus eas in tota naturali philosophia tractari, nec magis in calce, quàm in principio illius scientiae, tota enim in eis versatur, non sola ultima pars.

ultimate species, such as man, ox, ass, and all the others, and next move on to the heterogeneous parts and then to the homogenous, and from these to a treatment on the four elements, and finally arrive at consideration of first matter and of form, with which natural science, conveyed by Aristotle using compositive order, commences.

But those who hold that natural science can be conveyed using such resolutive order appear not to know what resolutive order is. For this (as they themselves also confess) is based on a notion of the end. Now the lowest species can, of course, be called the ultimate but not the end, because not everything ultimate is an end, as Aristotle teaches in the second book of the *Physics*.[29] But now for this issue to be understood, it has to be understood° that the whole of natural philosophy conveyed by Aristotle maintains the order that is from universal wholes to each [of the parts], just as he insists in the proem of the first book of the *Physics*.[30] For in the eight books of the *Physics* he first considers natural body accepted in its widest sense and makes its beginning-principles and affections clear. Afterward,[31] in the books *On the Heavens*, he takes into consideration simple body, which is the first species of natural body; then mixed body taken in a broad sense; next its species, in order, all the way to the ultimate [species]. Because of this, many, deceived by that which appears on first glance, believed that knowledge of the ultimate species is the ultimate end for the natural philosopher, since knowledge of genera appears to be inquired after for the sake of knowledge of species.

But this is not so. For while we might admit [for the sake of argument] that the species themselves are such an end that can establish resolutive order—which, of course, is not at all true—we nevertheless say that in natural philosophy as a whole they are treated no more at the end than at the beginning of that science, for the whole and not only the last part is concerned with them.

15

16

17 Genera namque non reperiuntur extra species suas, sed sunt ipsaemet species, quae sub ratione° communi considerantur, animal enim considerare quatenus est animal nihil aliud est considerare, quàm hominem, equum, bovem et alias species sub hac communi conceptione quatenus animalia sunt. Idcircò tractari in libris de Naturali auscultatione corpus naturale quà naturale est, nil aliud est, quàm tractari hominem, equum, ignem, aerem, aurum, aes et alias omnes species corporis naturalis sub hac communissima consideratione ut naturalia corpora sunt, similiter in libris de Ortu et interitu tractari de misto est tractari de ultimis speciebus quatenus sunt corpora mista. Quod idem de aliis inferioribus generibus dicendum est. Postrema omnium est earundem specierum tractatio secundùm proprias singularum naturas, quibus unaquaeque ab aliis discrepat. Hanc igitur ultimam tractationem si spectemus, ea est quidem ultima, non tamen est finis respectu praecedentium, quia illae quoque fuerunt de eisdem speciebus, licèt sub conceptibus communioribus, qui non minùs sunt per se digni cognitu, quàm proprii ipsarum specierum conceptus. Quòd si totam et perfectam specierum tractationem accipiamus, hoc est secundùm omnes ipsarum conceptus, haec quidem finis est philosophi naturalis, qui vult species corporis naturalis perfectè cognoscere, sed ipse non minùs ab his incipit, quàm in his desinit, quia ubique de his speciebus agitur secundùm diversos conceptus.

18 Hoc intellexit Averroes in sua praefatione in primum librum Physicorum, quando dixit intentionem philosophi naturalis esse cognoscere causas specierum sensilium et causas accidentium, quae sunt in eis, non enim intellexit proprias tantùm specierum causas et propria singularum accidentia, sed omnes causas et omnia accidentia ipsarum, etiam quae eis tribuuntur ratione superiorum conceptuum.

For genera are not found outside their species but are the spe- 17
cies themselves considered under a common account.° For to consider animal insofar as it is animal is to consider nothing other
than man, horse, ox, and the other species under this common
conception, that is, insofar as they are animals. Accordingly, for
natural body qua natural, to be treated in the *Nature Lectures* [i.e.,
the *Physics*] is nothing other than for man, horse, fire, air, gold,
bronze, and all the other species of natural body to be treated under this most common consideration, that is, as they are natural
bodies. Likewise in the books *On Coming to Be and Passing Away*:
for mixed [body] to be treated is for ultimate species to be treated
insofar as they are mixed bodies. The same thing has to be said
about the other lower genera. The final treatment of all is of the
same species according to the proper natures of each, by which
each is different from the others. If, therefore, we look at the last
treatment, it is, of course, the last, but it is not the end with regard to the preceding, because these were about the same species
also, though granted under concepts more general, which are no
less worthy of being known *per se* than are the proper concepts of
the species themselves. Now if we grasp the whole and perfect
treatment of the species, that is, according to all their concepts,
this, of course, is the end of the natural philosopher, who wants to
know perfectly the species of natural body; but he himself starts
no less from these than he finishes in them, because these species
are dealt with everywhere according to different concepts.

This [is what] Averroës understood in his preface to the first 18
book of the *Physics*,³² when he said that the intention of the natural philosopher is to know the causes of sensible species and the
causes of accidents that are in them. He did not understand [it to
be] only the proper causes of the species and the proper accidents
of each of them, but all the causes and all their accidents, even
those that are ascribed to them by reason of the higher concepts.

19   Sed utcumque se res habeat, species corporis naturalis non sunt talis finis, qualem intelligere° debemus cùm de ordine resolutivo loquentes dicimus eum esse à notione finis. Etenim finem aliquem extra nostram cognitionem et ab ea diversum intelligimus, qui cùm nondum sit, fieri tamen à nobis liberè operantibus potest.

20   Res autem naturales et aliae, quae in reliquis scientiis speculativis tractantur, aut semper sunt aut factae sunt vel fiunt à natura aut ab aliqua alia causa, non à nobis, quare in his contemplandis non habet locum ille progressus, qui vocatur à notione finis, sed sola earum cognitio finis noster est, quem quidem finem talem esse diximus, ut necessariò ordinem compositivum requirat, qui igitur dicunt scientiam naturalem tradi posse ordine resolutivo, decipiuntur et propriam huius ordinis naturam et conditionem ignorant.

: VIII :

Quòd neque ad scientiae speculativae
inventionem conferat ordo resolutivus.

1   Aliqui fuerunt, qui putaverunt scientiam naturalem alio quidem ordine, quàm compositivo, tradi non potuisse, tamen inventam fuisse ordine resolutivo, proinde resolutivum non ad ipsius traditionem, sed ad inventionem utilem extitisse; Aristoteles namque eam scientiam scripsit ordine compositivo, sed ad causarum absconditarum inventionem resolutione usus est et ab effectis posterioribus ad causas priores processit, nam in primo libro Physicorum argumento sumpto à generatione duxit nos in primae materiae

But howsoever this may be, species of natural body are not   19
such an end as we ought to mean° when, speaking about resolutive
order, we say that it is based on a notion of the end. For we understand some end outside our knowledge and different from it,
which although it does not yet exist, can nevertheless be made by
us, practicing freely.

Now natural things and other things that are treated in the rest   20
of the speculative sciences, either are always, or are made or happen by nature or by some other cause, not by us. And so in contemplating these things the progression that is described as based
on a notion of the end, has no place; knowledge alone of these
things is our end, and, of course, as we said, this is such an end
that it necessarily requires compositive order. Those, therefore,
who say that natural science can be conveyed using resolutive order are deceived and are ignorant of the nature and character
proper to this order.

: VIII :

That resolutive order also does not contribute
to discovery of speculative science.

There were some who held that of course natural science could   1
not be conveyed using any order other than compositive, but that
it was nevertheless discovered using resolutive order, and so the
resolutive was useful not to its conveyance, but to [its] discovery;
Aristotle wrote this science using compositive order, but he used
resolution for discovery of hidden causes, and he proceeded from
posterior effects to prior causes; for in the first book of the *Physics*,
in an argument taken from generation he led us into knowledge of

cognitionem, et in octavo libro ex aeterno motu demonstravit dari primum aeternum motorem. Sed hi, ut alii benè animadverterunt, in ambiguitatem manifestam lapsi sunt, cùm pro ordine methodum acceperint, quòd enim ad inventionem principiorum naturalium opus fuerit resolutione verum est, ea tamen est methodus resolutiva et demonstratio ab effectis, de qua postea loquemur, non est ordo resolutivus. Et miror quomodo non viderint ea, quae ferè ab omnibus dici solent in interpretatione prooemii primi libri Physicorum, tritum est enim et vulgatum omnium dictum eo in loco quòd aliud est ordo doctrinae, aliud via doctrinae, quae methodus propriè dicitur.

2   Alii verò fuerunt, qui in alio sensu dixerunt ordinem resolutivum fuisse utilem ad naturalis scientiae inventionem, illi enim, qui primi eam scientiam composuerunt, videntur primo loco ipsam hoc ordine invenisse, primùm quidem eis occurrerunt species ultimae corporum naturalium et desiderio ducti sunt cognoscendi ipsarum naturas et causas et affectiones, postea coeperunt proximas singulorum causas considerare et ab illis ad remotiores ascendere, donec tandem ad primas et remotissimas pervenerunt, à quibus exordium sumpsit traditio scientiae naturalis ordine compositivo. Inventio igitur scientiae naturalis praecessit per ordinem resolutivum, mox eam subsecuta est ipsius traditio per compositivum.

3   Nequaquàm tamen verum est id, quod isti comminiscuntur, primùm enim dubitare non absque ratione possumus an in illa prima rerum consideratione, quam ipsi inventionem scientiae vocant, is ordo servetur, quem arbitrantur, cùm enim in ultimis speciebus omnium superiorum generum naturae contineantur, proinde species ipsae nobis cognoscendae proponantur sub diversis conceptibus et magis et minùs communibus, quemadmodum diximus, rationi consentaneum esse videtur ut prius nobis offerantur et desiderium cognoscendi in nobis excitent sub conceptu

first matter, and in the eighth book, from eternal motion he demonstrated that there is a first, eternal mover. But they, as others
well noted, lapsed into manifest ambiguity, since they accepted
method for order. For it is true that resolution was needed for
discovery of natural beginning-principles. But this is resolutive
method and demonstration from effects, about which we will
speak below; it is not resolutive order. And I wonder how they did
not see that which is normally said by nearly everyone in [their]
interpretation of the proem of the first book of the *Physics*.³³ For
the dictum in that passage, that there is on the one hand order of
teaching and on the other the way of teaching—what is properly
called method—is familiar and commonplace to everyone.

And there were others who said resolutive order is useful for   2
discovery of natural science in another sense. For those who first
composed this science appear to have discovered it in the first
place using this order: First, of course, they encountered the ultimate species of natural bodies, and they were led by the desire for
knowing the natures, causes, and affections of these. Afterward
they began to consider the proximate causes of each and to ascend
from them to ones more remote, until finally they came through to
the first and most remote [causes], from which the conveying of
natural science using compositive order took [its] beginning. Discovery of natural science, therefore, first went by means of resolutive order; conveying it followed next by means of compositive.

But what they alleged is in no way true. For first we can doubt,   3
not without reason, whether, in that first consideration of things
that they call discovery of the science, the order that they think is
maintained, [is]. For since the natures of all the higher genera are
contained in the ultimate species, and so the very species to be
known to us are set out under different concepts, both more and
less general, just as we said, it appears to agree with reason that
they would be presented to us first and excite in us the desire for
knowing [them] under a more general concept, since that which is

communiore, siquidem id, quod magis universale est, prius et facilius à nobis cognoscitur, quàm minùs universale, idque tam secundùm distinctae quàm secundùm confusae cognitionis ordinem, prius enim cognovimus equum ut corpus, quàm ut animal et prius ut animal, quàm ut equum à bove et ab asino distinctum, facilius enim ea, quae multis communia sunt, inspicimus et cognoscimus, quàm quae singulorum sunt propria. Eo igitur ordine, quo res primùm à nobis confusè cognitae fuerunt, credendum est eas desiderium nostrum distinctae cognitionis excitare. Atqui si corpora naturalia primùm nos movent ut corpora, sequitur nos primùm excitari ad indagandam corporis naturam, deinde animalis et hominis, corporis autem naturam quà corpus est investigare est corporis principia quaerere, at principia corporis naturalis quà corpus naturale est sunt prima omnium rerum naturalium principia, de quibus agitur in libris de Naturali auscultatione. Videtur itaque haec scientiae inventio procedere ordine compositivo, non resolutivo.

4   Verùm ut eis condonemus eum ordinem tunc servari, quem ipsi imaginantur, id tamen certissimum nobis esse debet eam non posse scientiae naturalis inventionem appellari. Quid enim, quaeso, est scientia naturalis, nisi perfecta et distincta rerum naturalium per suas causas cognitio? Confusa namque et imperfecta cognitio scientia dici non potest. Videamus igitur an per hunc ordinem scientia naturalis inventa sit, dicunt isti prius offerri cognoscendas ultimas species, deinde earum causas proximas, postea remotiores, tandem primas ac remotissimas, in quo processu si intelligant nos à cognitione specierum et ultimorum effectuum venire in cognitionem proximarum causarum et ex his duci in cognitionem causarum remotiorum, in ambiguitatem incidunt, methodus enim ea est, non ordo, illationem enim habet ignoti ex noto. Ergo si ordinem, non methodum dicere debeant, necesse est ut ita intelligant,

more universal is known by us earlier and more easily than the less universal, and this as much according to the order of distinct [knowledge] as according to the order of confused knowledge. For we knew horse first as body, then as animal, and first as animal, then as horse distinct from ox and ass. For we observe and know more easily those things that are common to many rather than those that are proper to each. In that order, therefore, in which things were first known by us confusedly, it has to be believed, they excite our desire for distinct knowledge. But yet, if natural bodies first move us as bodies, it follows that we are first excited to track down the nature of body, and then of animal and of man. To investigate the nature of body qua body, however, is to inquire after the beginning-principles of body; but the beginning-principles of natural body qua natural body, are the first beginning-principles of all natural things; these are dealt with in the *Nature Lectures* [i.e., the *Physics*]. And so it appears that this discovery of the science proceeds using compositive, not resolutive, order.

But even if we grant them that this order they imagine is then 4 maintained, nevertheless it ought to be most certain to us that this cannot be called discovery of natural science. For what, I pray, is natural science except the perfect and distinct knowledge of natural things by means of their causes? Surely confused and imperfect knowledge cannot be said to be scientific knowledge. Let us see, therefore, whether natural science is discovered by means of this order. They say that the ultimate species to be known are presented first, and then their proximate causes, and afterward the more remote, and finally the first and most remote. If in this procedure they understand that we come into knowledge of proximate causes from knowledge of species and ultimate effects, and are led from these into knowledge of more remote causes, they fall into ambiguity.[34] For this is method, not order, for it has an inference from known to unknown. Therefore if they ought to say order, not method, it is necessary that they understand that first,

prius quidem ipsarum specierum cognitionem quaeri ac de iis prius tractari; postea verò agi de proximis ipsarum causis et earum cognitionem quaeri; deinde de remotis; demum de remotissimis principiis. Sed haec nullo pacto potest vocari scientia neque scientiae naturalis invento, quia tota illa cognitio confusa est. Dum enim prima et remotissima principia ignoramus, nullum aliorum possumus perfectè cognoscere, qualis igitur erit tractatio de homine vel de equo, dum remotiores omnes causae ignorantur? Certè confusa et rudis.

5   Praeterea in illa prima rerum consideratione, quam isti inventionem scientiae vocant, cognoscimusne in ipsis rebus aliquid an nihil? Si nihil, igitur per eam nihil scimus, quare nullius scientiae inventio est. Si aliquid, quodnam est illud? Ut quando primo loco hominem et alias species consideramus, quidnam in homine cognoscimus? Substantia quidem hominis ex forma et materia constat, nempè ex anima, et corpore, attamen non potest nobis esse cognitum quòd cognitio naturae humanae consistat in cognitione materiae et formae, quia[30] ex materia et forma constare, non est hominis proprium, neque ipsi competit quatenus est homo, sed corpori naturali quatenus corpus naturale est. Dum igitur ignoramus hanc esse omnis corporis naturalis essentialem conditionem, non possumus cognoscere quòd ad humanam naturam perfectè cognoscendam requiratur cognitio materiae et formae. Sed neque corpus et anima cognosci possunt dum formae naturam communiter ignoramus, necnon primam materiam, ex qua omnes aliae secundae materiae constitute sunt.

6   Alia quoque multa absurda hanc sententiam consecuntur, quae quisque considerare potest.

knowledge of the species themselves is inquired after and they [i.e., the species] are treated first, and afterward their proximate causes are treated and knowledge of them is inquired after, then the remote, and lastly the most remote beginning-principles. But this can in no way be called scientific knowledge or discovery of natural science, because as a whole that knowledge is confused. For when we are ignorant of the first and most remote beginning-principles, we can know none of the others perfectly. Of what sort, therefore, will a treatment on man or horse be, when all remote causes are unknown? Certainly confused and crude.

Moreover, in that first consideration of things that they call 5 discovery of the science, do we know something in the things themselves, or nothing? If nothing, then by means of it we know nothing scientifically, and so there is discovery of no science. If something, what is it? As when we consider in the first place man and other species, what do we know in man? Of course, the substance of man is composed out of form and matter, that is, of soul and body. It nevertheless cannot be known to us that knowledge of human nature consists in knowledge of matter and form, because to be composed out of matter and form is not proper to man; this does not appertain to him insofar as he is man, but to natural body insofar as it is natural body. When, therefore, we are ignorant that this is an essential characteristic of every natural body, we cannot know that knowledge of matter and form is required for knowing human nature perfectly. But neither can body and soul be known when we are ignorant of the nature of form in general and of first matter, from which all other second matters are constituted.

Many other absurdities ensue from this position also; anyone 6 can consider them.

7   Certum itaque atque compertum esse debet, quisquis sit ille ordo, qui in illa meditatione servatur, eum penitus ineptum esse non modò ad scientiae naturalis traditionem, sed etiam ad eiusdem inventionem.

8   Est etiam maximè rationi consonum eodem prorsus ordine traditam esse à philosophis scientiam naturalem, quo ab eis inventa prius fuit, nam si solo ordine compositivo scientia naturalis tradi potuit, quia perfecta rerum naturalium cognitio alio ordine non potest comparari, sequitur primos quoque huius scientiae inventores alio ordine perfectam rerum naturalium scientiam invenire et consequi non potuisse. Quod etiam de methodo asserendum est, qua enim methodo Aristoteles perfectam nobis tradidit scientiam naturalem, necesse est eadem ipsum quoque ad rerum naturalium perfectam cognitionem ductum fuisse et eadem fuisse inventam scientiam naturalem. Fraudasset quippe nos Aristoteles vel quisquis alius naturalis scientiae primus inventor fuit, si eo ordine eaque methodo dimissis, quibus ipse utens scientiam naturalem invenit et rerum naturalium cognitionem optimam consecutus est, alium ordinem et aliam methodum in ea nobis tradenda servasset, quibus vel non ita perfectam vel difficilius rerum naturalium cognitionem adipisceremur.

9   Quod autem de naturali scientia diximus, id de omnibus aliis contemplativis scientiis existimandum est, eadem enim in omnibus ratio viget.

10   Manifestum est igitur, scientiis contemplativis tum tradendis tum inveniendis solùm convenire ordinem compositivum, eadem enim sese offert rerum cognoscendarum natura et illi, qui contemplando ac laborando earum scientiam invenire vult et illi, qui eam aliis tradere constituit.

And so it ought to be certain and assuredly known that what- 7
ever that order is that is maintained in that meditation, it is completely inapt, not only for the conveying of natural science but also
for the discovery of the same.

Also, it is most consonant with reason that natural science be 8
conveyed by philosophers using entirely the same order in which it
was first discovered by them. For if natural science could be conveyed using resolutive order alone, then perfect knowledge of natural things could not be procured using another order. It follows
too that the first discoverers of this science could not discover and
gain perfect scientific knowledge of natural things using another
order. This has to be asserted about method also. For by whatever
method Aristotle conveyed perfect natural science to us, it is necessary that he was also led to perfect knowledge of natural things
by the same [method] and natural science was discovered by the
same [method]. Of course, Aristotle, or whoever else was the first
discoverer of natural science, would have deceived us, if, putting
aside that order and that method which he used to discover natural science and gain optimal knowledge of natural things, he had
maintained another order and another method in conveying things
to us, by which we would either not obtain such perfect knowledge of natural things or do so with much difficulty.

Moreover, what we said about natural science should be judged 9
so of all other contemplative sciences, for the same reasoning applies in all [of them].

It is, therefore, manifest that compositive order alone is appro- 10
priate both for conveying and discovering contemplative sciences.
For the nature of the things to be known is presented in the same
way both to those who by contemplating and laboring want to
discover the science of them and to those who decide to convey it
to others.

11   Illa autem, quam aliqui inventionem vocabant, certum est neque inventionem scientiae neque traditionem esse, sed prae-
cedentem quandam considerationem et praemeditationem, dum
rerum cognoscendarum desiderio atque amore movemur, antequàm ipsarum cognitionem indagare incipiamus.

: IX :

Quòd artes et disciplinae aliae omnes praeter
contemplativas solo ordine resolutivo tradi possint.

1   Contra verò se res habet in artibus ac omnibus aliis disciplinis,
quae non scientiam quaerunt, sed finem aliquem à nobis agendum
vel efficiendum. Hae namque alio ordine tradi non possunt, quàm
resolutivo. Ratio autem ex ipsa talium disciplinarum natura hunc
in modum colligitur.

2   Certum est duo in unaquaque arte esse, quae in considerationem cadunt, unum quidem et praecipuum est[31] finis ipse ac-
quirendus sive producendus à nobis, alterum verò principia, ex
quibus ipsum producamus, quae possunt etiam appellari[32] media
ad finem illum ducentia, ut in arte aedificatoria finis est domus
ipsa, principia verò lapides, lateres, ligna et alia eiusmodi. Domus
enim si iam existens, aut facta à natura vel ab aliquo alio agente
nobis proponeretur, nullis certè lapidibus, nullave alia materia pro
ea construenda egeremus, sed absque ullis mediis fine optato potiremur. Verùm quia domus neque est neque fit ab alio, sed à nobis,
ideò materia indigemus, ex qua ipsam efficiamus, ars enim solam
artificialem formam gignere in materia potest, materiam generare
nequit, sed eam à natura genitam accepit.[33]

But now, that which some were calling discovery is certainly   II
neither the discovery nor the conveying of a science, but is some
sort of preceding consideration and premeditation, when we are
moved by desire and love for the things to be known, before we
start to track down the knowledge of them.

: IX :

That arts and all other disciplines besides the contemplative
can be conveyed only using resolutive order.

But it truly is otherwise in arts and all other disciplines that do   I
not inquire after scientific knowledge but [seek] some end, to be
done or effected by us. For these cannot be conveyed using any
order other than resolutive. And the reason is gathered in the following way from the very nature of such disciplines.

Certainly, in every art there are two things that fall under con-   2
sideration: one, and the principal one, is the very end to be
achieved or produced by us; the other, the beginning-principles
from which we produce that [end] itself and which can thus also
be called the means leading to that end. As in the art of building,
the house itself is the end, and the stones, bricks, wood, and other
things of this type, the beginning-principles. For if a house already
existing, either made by nature or by some other agent, were set
out before us, certainly we would need no stones or any other
material to construct it; instead, without any means we could get
hold of the desired end. But because a house neither is nor comes
to be by another [agent], but by us, we are, therefore, in need of
matter from which we may effect it. For art can engender only an
artificial form in matter; it cannot generate matter, but accepts
that [matter] engendered by nature.

3   Ordo igitur generationis, ac operationis in omni arte est ex necessitate compositivus, nempè à principiis ad finem, à simplicibus ad composita, conditio namque est perpetua et lex necessaria omnis generationis sive à natura sive ab arte sive à quovis alio agente factae, ut ordinem servet compositivum. Ideò Aristoteles in contextu 23 septimi Metaphysicorum loquens de ordine artis operantis solum ei tribuit compositivum, qui est à principiis.

4   Sed ordo artis docentis est ex necessitate contrarius ordini artis operantis, nam finis ipse nunquàm proponitur ignotus, nunquàm quaeritur, nec unquàm de fine fit aliqua consultatio, ut docet Aristoteles in tertio capite tertii libri de Moribus. Sed semper proponitur finis aliquis certus consequendus, ut finis exercitus est victoria, de qua nunquàm fit consultatio an quaerenda sit, necne; sed certum et constitutum in omnium militum animo est vincendum esse; disceptatio autem tota ac tota consultatio est de mediis, quibus victoria acquiri possit. Obsidentium civitatem finis certus est expugnatio, consultatio autem est de modo et de mediis ad eius expugnationem conferentibus.

5   Cùm igitur in omni arte finis notus proponatur, si media quoque ad illum ducentia et principia, ex quibus constituendus est, per se nota sint, nulla arte docente opus est, sed sola arte operante. Artis enim doctrina ad operationem dirigitur, ideò ubi omnia sunt nota tum principia tum finis, tota disciplina supervacanea est, iam enim quisque finem illum consequi desiderans potest per se ipsum aggredi ex principiis notis operationem. Si est igitur ars aliqua docens, aliquid esse ignotum oportet, quod discendum proponatur, is non potest esse finis, erunt igitur principia, è quibus

The order of generation and practical activity in every art, 3
therefore, is out of necessity compositive, that is, from beginning-
principles to end, from simples to composites. For it is a perpetual
characteristic and necessary law of all generation made either by
nature, by art, or by any other agent that compositive order be
maintained. Because of this, Aristotle in text no. 23 of the seventh
[book] of the *Metaphysics*,[35] speaking about the order of an art
being practiced, ascribes to it only compositive, which is from
beginning-principles.

But the order of an art being taught is out of necessity the con- 4
trary of the order of an art being practiced. For the end itself is
never set out as unknown, it is never inquired after, and no deliberation about the end ever occurs, as Aristotle teaches in the third
chapter of the third book of the *Ethics*.[36] But some definite end to
be gained is always set out, as [for example] an army's end is victory. About this there is never deliberation on whether it is to be
sought after or not; that they have to win is definite and decided
in the soul of all the soldiers, and the whole debate and the whole
deliberation is about the means by which victory can be achieved.
The definite end of those besieging a city is to take it; the deliberation is about how and about the means that contributes to taking it.

Since in every art, therefore, a known end is set out, if the 5
means leading to it and the beginning-principles from which it is
to be constituted are also known *per se*, there is no need for an art
to be taught, only for an art to be practiced. For the teaching of art
is directed to practical activity. And so when all are known, both
the beginning-principles and the end, the whole discipline is superfluous. For then anyone desiring to gain that end can by his
own means undertake the activity from the known beginning-
principles. If, therefore, there is some art being taught, something
must be unknown, that which is set out to be learned. This cannot
be the end; therefore, it is the beginning-principles from which it

constituendus est. Atqui ab ignotis ordiri non possumus, sed semper à notis, ut anteà universè de omni ordine demonstravimus et ut asserit Aristoteles in quarto capite primi libri de Moribus. À fine igitur in omni arte docente inchoandum est et ad principiorum inventionem progrediendum.

6   Hoc idem de omni alia disciplina confiteri cogimur, cuius natura non in contemplando, sed in operando consistat, eadem enim in omnibus ratio locum habet et alio ordine tradi non possunt, quàm resolutivo. Quod mihi quidem videtur rei naturam considerando ita perspicuum esse, ut mirandum profectò sit quomodo alii in aliam sententiam inciderint, artis enim docentis nullus alius est scopus, quàm media invenire, quae ad propositum finem perducant, quae si nota essent, tota artis doctrina supervacanea foret; cùm autem sint ignota, ex iis progredi operando non possumus, nisi beneficio artis docentis illa per resolutionem finis prius invenerimus et cognoverimus. Quoniam enim non simpliciter eorum cognitio quaeritur, sed solùm ut ad talem finem conferentium et is proponitur notus, sequitur ea ex sola collatione° cum ipso fine posse manifesta fieri, ideò in eorum cognitionem non pervenimus nisi ex notione finis. Ex ipsa igitur talium disciplinarum tradendarum natura ortum habet necessariò ipse ordo resolutivus. Est igitur ex necessitate contrarius ordo artis docentis ordini artis operantis, quia necesse est in iis desinere artis doctrinam, à quibus operatio incipit; et ab iis doctrinae exordium sumi, in quibus operatio desinit.

[i.e., the end] has to be constituted. Yet we cannot begin with the unknown, always with the known, as we demonstrated earlier [and] universally about every order and as Aristotle asserts in the fourth chapter of the first book of the *Ethics*.[37] In every art being taught, therefore, one has to start with the end and has to progress to discovery of beginning-principles.

We are forced to confess this same thing about every other dis- 6 cipline whose nature consists not in contemplating but in practicing. For the same reasoning has a place in all of them, and they cannot be conveyed using an order other than resolutive. Of course, this appears to me to be so plain, in considering the nature of the issue, that it surely has to be wondered how others could have fallen into another position. For the goal of an art being taught is none other than to discover the means that lead through to a given end. If these were known, the whole teaching of the art would be superfluous, but since they are unknown, we cannot progress from them in practicing, unless, thanks to the art being taught, we have first discovered and come to know them by means of resolution of the end. For since knowledge of them is not inquired after absolutely, but only as contributing to such an end, and this is already set out [as] known, it follows that they can become manifest just from a comparison° with the end itself. Therefore we do not come through into knowledge of them except from a notion of the end. Resolutive order itself, therefore, necessarily came to be from the very nature of those disciplines being conveyed. The order of the art as taught, therefore, is out of necessity contrary to the order of the art as practiced, because it is necessary that the teaching of the art finish in those things from which the practical activity starts and the beginning of teaching be taken from those things in which the practical activity finishes.

: X :

Confirmatio dictae sententiae ex Aristotelis, Galeni, Avicennae et Averrois dictis.

1   Sententiam hanc apud Aristotelem legimus in contextu 23 septimi
Metaphysicorum saepe à nobis memorato, ubi veritatem pulchrè
declarat utens exemplo artis aedificatoriae, inquit enim in lapidibus ac primis elementis desinere artis eius doctrinam et ab eisdem
incipere operationem; quemadmodum ab eodem fine doctrinae
initium sumitur, in quo acquisito desinit operatio. Unde colligit,
quòd modo quodam domus domum facit, nempè domus praecognita, domus mentalis domum materialem. Cùm enim prius à
domo praecognita ad elementa cognoscenda, deinde ab his inventis
ad domum realem efficiendam progrediamur, manifestum est, totum simul progressum doctrinae et operationis à domo esse ad
domum. Hoc modo censuit Aristoteles locum habere in artibus
utrumque ordinem, in doctrina quidem resolutivum solum, in
operatione verò compositivum.

2   Quòd si quis arti illi aedificatoriae tradendae, aut scribendae
ordinem compositivum aptare contenderet, quem eundem in
operatione servari necesse est, ipsemet si rationis compos esset, ex
ipsa re facilè intelligeret quàm vanum ac inutile opus aggrederetur,
nec enim quid quaereret nec ex quibus ad illud progrederetur effari
potis esset.

: X :

*Confirmation of the said position, from what is said
by Aristotle, Galen, Avicenna, and Averroës.*

We read this position in Aristotle in text no. 23 of the seventh   1
[book] of the *Metaphysics*,[38] so often referred to by us, where he
beautifully makes the truth clear, using the example of the art of
building. For he says teaching of the art finishes in stones and first
elements, and practical activity starts from the same, just as the
start of teaching is taken from the same end in the acquisition of
which the practical activity finishes. From this he gathers that in
some way house makes house, that is, the house known beforehand, the mental house, [makes] the material house. For since we
first progress from the house known beforehand to the elements to
be known, and then from these discovered things to the real house
that is to be effected, it is manifest that the whole progression of
teaching and practical activity together is from house to house. In
this way Aristotle deemed that each order has a place in the arts —
the resolutive alone in teaching, but the compositive in practical
activity.

   Now if someone contended that compositive order applies to   2
conveying or writing the art of building, the same [order] that it is
necessary to maintain in practical activity, then if he were in possession of reason, he would easily understand from the matter itself what a vain and useless work he has undertaken. For he would
not be able to describe either what he was inquiring after or what
he was progressing toward it from.

3   Praeterea in quinto capite sexti libri de Moribus et in capite octavo septimi inquit Aristoteles finem in actionibus esse principium, sicut in mathematicis suppositiones, ut enim in mathematicis suppositiones demonstrari non possunt, sed eas praecognosci necesse est, ut reliqua ex ipsis manifesta fiant, ita in activis disciplinis debet finis esse totius cognitionis principium et in artibus omnibus similiter. Propterea tradidit Aristoteles moralem[34] facultatem ordine resolutivo, à felicitate namque exordium sumpsit, quae est operatio ex virtute proveniens et est ultimus omnium agendorum finis, hanc in primis dari constituit, deinde ipsius definitionem assignat, postea in secundo libro incipit agere de virtute. De Aristotelis igitur sententia dubii esse non possumus, ex eius enim lectione vel caecus inspicere potest quid ipse senserit.

4   Hoc idem et Galenus fassus est in libro de Artis medicae constitutione, nam in primo et ultimo et penultimo capitibus asserit artem medicam constitui à notione finis, quia est de numero artium effectricium, quae omnes à finis sui notione constituendae sunt. Quare non arbitrarium vult in artibus esse ordinem resolutivum, ut alio quoque ordine pro scribentis arbitrio tradi possint, sed necessarium, ut nullum alium constitutio alicuius artis recipiat, hancque dicit veterum philosophorum sententiam extitisse. Itaque in eo libro declarat modum tradendi medicam artem ordine resolutivo. Sunt etiam qui dicant Galenum ibi non solùm modum ostendere tradendi artem medicam, sed ipsam quoque artem eo ordine tradere. Sed qua ducti ratione hoc asserant, non video, cùm ex eius libri lectione falsitas huius opinionis facilè deprehendatur.

Moreover, in the fifth chapter of the sixth book of the *Ethics*, 3
and in the eighth chapter of the seventh [book],[39] Aristotle says
that, in actions, the end is the beginning (*principium*), just as in
mathematics, the suppositions are. For as in mathematics the suppositions cannot be demonstrated—it is necessary they be known
beforehand, so that from them the rest may become manifest—so
in active disciplines, the end ought to be the beginning of the
whole [body of] knowledge, and similarly in all the arts. And so
Aristotle conveyed the moral branch of learning using resolutive
order. He took the beginning from felicity, which is practical activity arising out of virtue and is the ultimate end of all things to be
done. He establishes in the first place that there is [such a thing]
and then assigns it a definition;[40] afterward, in the second book,
he starts to deal with virtue. We cannot, therefore, be in doubt
about Aristotle's position. For from a reading of it even a blind
man can observe what he thought.

And Galen acknowledged the same in the book *On the Constitu-* 4
tion of the Medical Art. For in the first, last, and penultimate chapters[41] he asserts that the medical art is constituted from a notion
of the end, because it is numbered among the effective arts, all of
which are to be constituted from a notion of their own end.
Therefore he wants resolutive order not to be discretionary in the
arts, such that they could be conveyed using another order too by
the choice of the one writing, but necessary, such that the constitution of any art receives no other [order]. And he says this was
the position of ancient philosophers. And so in this book he
makes the way of conveying medical art using resolutive order
clear. Still, there are those who say Galen there not only shows the
way of conveying the medical art, but also conveys the art itself
using that order. But I do not see by what reason they are led to
assert this, since the falsity of this opinion is easily apprehended
by a reading of the book.

5 Avicenna quoque artem medicam scripsit ordine resolutivo, non compositivo, ut medici arbitrantur, idque nos deinceps apertissimè demonstrabimus. Aut enim Galenus et Avicenna huiusce rei veritatem non igoraverunt, aut saltem ab ipsa artis natura, quae alium ordinem non patitur, ad eum servandum tracti atque coacti fuere.

6 Galenus quidem in libro de Artis medicae constitutione videtur veritatem eiusque rationem non ignorasse, attamen dum artem suam medicinalem respicio, nescio quid de ipso dicam, nam in ea similiter ordo servatur resolutivus, ut mox demonstrabimus, ipse tamen alium ordinem esse putat, nempè definitivum, quem antea reiecimus. Propterea an liber ille Galeni fuerit, necne, non possum non magnopere dubitare.

7 Averroes autem in suo libro, qui inscribitur Colliget, non modò artem medicam tradit ordine resolutivo, sed et eius ordinis conditiones ita egregiè declarat, ut in illius artis natura cognoscenda et in ea artificiosissimè disponenda significet se exquisita logicae cognitione praeditum fuisse et in ea tam Galeno quàm Avicennae longè esse praeferendum. Potest enim liber ille vocari idea ordinis resolutivi, cùm eum exactè tum servet tum declaret ibi Averroes, ut mox ostendemus.

: XI :

In quo declaratur in artis medicae
traditione conditio ordinis resolutivi.

1 Quoniam autem ex artis medicae traditione pulcherrimum exemplum sumi potest ordinis resolutivi, non erit ab re si eam aliquantum consideremus, quanquàm enim medici non sumus, neque de

Avicenna too wrote [his] medical art[42] using resolutive, not 5
compositive order as the physicians think; we will demonstrate
this very plainly in what follows. For either Galen and Avicenna
were not ignorant of the truth of this issue or at least they were
dragged and forced to maintain it by the art's very nature, which
cannot tolerate another order.

Galen, of course, in the book *On the Constitution of the Medical* 6
Art, does not appear to have been ignorant of the truth and its
reason. Nevertheless, when I give regard to his *Art of Medicine*, I do
not know what I would say about it. For in it resolutive order is
similarly maintained, as we will soon demonstrate. Nevertheless,
he holds that it is another order, namely definitive; we earlier rejected this. Therefore I cannot but doubt greatly whether or not
this book was Galen's.

Now Averroës, in his book that is titled *Colliget*,[43] not only con- 7
veys medical art using resolutive order, but also clarifies the characteristics of this order so very well that he indicates that, in
knowing the nature of that art and in disposing it so very skillfully,
he was endowed with a fine knowledge of logic and in this is by far
to be preferred to Galen as well as to Avicenna. For that book can
be called the [very] idea of resolutive order, since Averroës there
both maintains and clarifies it exactly, as we will soon show.

: XI :

In which a characteristic of resolutive order in
the conveying of medical art is made clear.

Now since a most beautiful example of resolutive order can be 1
taken from the conveying of medical art, it will not be beside the
point if we consider it a little. For although we are not physicians

re medica, sed[35] de rebus logicis in praesentia nobis sermo institutus est; attamen si eius artis exemplo quid sit ordo resolutivus et quomodo servetur docuerimus, non parvum certè[36] operae pretium fecerimus. Logica enim omnium disciplinarum instrumentum est, et quemadmodum Aristoteles ex ipsis disciplinis, quas sapientes viri tradiderant, logicam artem desumpsit, atque construxit; ita nobis quoque licebit ex aliis disciplinis, à quibus logica ortum habuit, rerum logicarum declarationem depromere, tanquàm per logicam in usu positam praecepta artis logicae explicantibus.

2   Et quemadmodum Galenus in libro de Artis medicae constitutione unius morbi exemplo proposito totius artis constitutionem declarat, ita et nos sumpto medicinae exemplo ordinem, quo artes omnes tradendae sunt, explicare commodè poterimus.

3   Hac autem in re hunc ordinem servandum esse duximus, prius enim artis huius naturam contemplantes considerabimus ratione duce quem ordinem traditio illius artis requirat et undenam sumendum sit cognitionis eius initium, deinde eundem ordinem in Averrois et Avicennae et Galeni traditionibus servatum esse ostendemus.

4   Illud in primis à nemine literato viro et in arte medica mediocriter erudito ignoratum fuisse arbitror, artem illam necessariò postulare ut à cognitione humani corporis exordium doctrinae sumatur, nempè ut omnes ipsius partes tam homogeneae quàm heterogeneae et singularum naturae ac temperies, necnon et officia et operationes ante omnia cognoscantur. Hoc quidem medicis incognitum fuisse non videtur, at huius rationem an benè noverint nescio. Ratio autem duplex est et utramque apertè declarat Averroes in suo Colliget; utramque etiam visus est significare Galenus in capite undecimo libri sui de Artis medicae constitutione. Una

and our discourse at the present time is intended by us to be about logical and not about medical issues, nevertheless if by the example of this art we taught what resolutive order is and how it is maintained, we would certainly have done something of no little value. For logic is an instrument of all disciplines, and just as Aristotle drew up and constructed the art of logic from the very disciplines that wise men had conveyed, so too it shall be permitted to us to draw out a clarification of logical issues from the other disciplines from which logic came to be, as it were, explicating precepts of the art of logic by means of logic put into use.

And in the same way that Galen in the book *On the Constitution* 2 *of the Medical Art*, by setting out the example of one single disease, clarified the constitution of the whole art, so also we will, by taking up the example of medicine, be able to advantageously explicate the order in which all the arts are to be conveyed.

Now in this issue we have come to think the following order 3 has to be maintained: First, contemplating the nature of this art and led by reason, we will consider what order the conveying of this art requires and from where the beginning of knowledge of it has to be taken. Then we will show that the same order is maintained in the conveying of Averroës, Avicenna, and Galen.

I think that no man who was lettered and moderately learned 4 in medical art could in the first place be ignorant of what that art necessarily demands — that the beginning of teaching be taken from knowledge of the human body, that is, that all its parts, homogeneous as well as heterogeneous, and the natures and tempers[44] of each, and both [their] functions and activities be known before all else. It appears, of course, that this was not unknown to physicians, but I do not know whether they well knew the reason for it. In reality there are two reasons, and Averroës clarifies both plainly in his *Colliget*,[45] and Galen too seems to indicate both in the eleventh chapter of his book *On the Constitution of the Medical Art*.[46] One is what we touched on in our book *On the Nature of*

est, quam tetigimus in libro nostro de Natura Logicae, dum de libro Categoriarum sermonem haberemus. Ob eandem enim rationem, ob quam voluit Aristoteles in principio logicae de decem summis generibus agere, debet etiam medicus artem docens à partibus humani corporis tractationem exordiri. Humanum enim corpus est subiectum, in quo medicus vult efficere vel conservare sanitatem, nullus autem artifex potest operari in subiecto penitus incognito, sed tantam eius debet habere praecognitionem, quantam ars illa et opus eius exposcit, quemadmodum testatur Aristoteles in 26 contextu secundi libri Physicorum; et clarius in capite ultimo primi libri de Moribus, ubi hoc ipso exemplo utitur. Sicuti medicus oculum curaturus debet ante omnia aliquam oculi cognitionem habere, quam illa curatio postulat, ita moralis philosophus animam sanaturus debet in primis aliquam animae, saltem levem, habere notitiam° et partes animae pingui Minerva cognoscere.

5   Toti igitur arti medicae tam operanti quàm docenti necessaria omnino est praecognitio humani corporis et omnium eius partium cum propriis singularum officiis, et operationibus. Quomodo enim vitia et morbos humani corporis et partium eius rectè curabit medicus, nisi partes ipsas cognoscat? Ut rectè Galenus dicit in capite II libri de Artis medicae constitutione.

6   Hac ratione cogitur omnis ars docens tradere ante omnia proprii subiecti cognitionem aliquam, postea in eo rationem et modum operandi docere, nisi contingat ipsum subiectum artis ita esse per se conspicuum, ut à nemine tanta eius notitia° non habeatur, quantam ars illa requirit. Hoc enim si eveniat, spernitur huiusmodi tractatio, quia supervacaneis non est immorandum. Tale autem profectò non est humanum corpus et eius partes, harum enim quamplurimae internae et absconditae sunt; quae verò sunt externae, pingui Minerva ab hominibus vulgaribus noscuntur,° harum

Logic when we discoursed upon the book the *Categories*.[47] For by the same reasoning by which Aristotle wanted in the beginning of logic to deal with the ten highest genera, so also the physician teaching the art ought to begin [his] treatment with parts of the human body. For the human body is the subject in which the physician wants to effect or conserve health. And no practitioner can work on a completely unknown subject; he ought instead to have as much prior knowledge of it as the art and its work calls for, just as Aristotle attests in text no. 26 of the second book of the *Physics*[48] and even more clearly in the last chapter of the first book of the *Ethics*,[49] where [he] uses this very example: Just as a physician who would cure an eye ought to have before all else some knowledge of the eye, [knowledge] that the cure demands, so the moral philosopher who would heal the soul ought in the first place to have at least some light knowledge° of the soul and to know the parts of the soul, even if just coarsely.

In the whole medical art, therefore, in practicing as much as in   5 teaching, prior knowledge of the human body and all of its parts, with the proper functions and activities of each, is altogether necessary. For how will a physician correctly cure impediments and diseases of the human body and of its parts unless he knows the parts themselves? — as Galen correctly says in chapter 11 of the book *On the Constitution of the Medical Art*.[50]

For this reason every art being taught is forced, before all else,   6 to convey some knowledge of its proper subject, and afterward to teach the reason and the way of practicing in it — unless it happens that the subject itself is so plainly apparent *per se* that everyone has as much knowledge° of it as the art requires. And if this happens, a treatment of this type is left aside, because there should be no tarrying over superfluous things. But surely the human body and its parts are not such. For a great many of them are internal and hidden, while those that are external are known,° if just coarsely,

enim naturas ac temperies, quas cognoscere medico necessarium est, cuncti praeter eruditos viros ignorant. Prior itaque ratio haec[37] est, quae à subiecto medicae artis desumitur.

7   Altera est ratio, quae sumitur à fine, maximè quidem consideratione digna, sed aliis, ut videtur, prorsus incognita, haec enim medicum omnino cogit in artis medicae traditione ab humano corpore, ac eius partibus declarandis auspicari. Nam secus facere minimè potest, si est usurus ordine resolutivo à notione finis. Etenim praecognoscendus est finis duplici praecognitione, quemadmodum et de subiecto scientiae docet Aristoteles in primo libro Posteriorum Analyticorum, ut enim subiectum in scientia est praecognoscendum et quòd sit et quid sit; ita finis in arte praenosci° debet tum quòd esse seu fieri possit, tum quid sit. Nisi enim esse posset, frustra quererentur media pro eius generatione. Si verò quid sit ignoremus, nulla ratione invenire illa media possumus principio ad eorum inventionem idoneo destituti, quod unum est definitio finis, ex hac enim una duci possumus ad mediorum seu principiorum cognitionem. Docet hoc Aristoteles in contextu illo 23 septimi Metaphysicorum dicens artium omnium doctrinam fieri à notione finis, nilque aliud esse notionem finis, quàm ipsius finis definitionem.

8   Est autem finis artis medicae sanitas tum conservanda, si adsit; tum recuperanda, si lapsa fuerit. Debet igitur medicus ante omnia praenoscere° sanitatem amitti et recuperari posse, quod quidem notissimum omnibus est: deinde etiam quid sit sanitas et in quo consistat, hoc enim si ignoret, quomodo remedia pro sanitate tuenda vel recuperanda poterit indagare? Sanitas autem nihil aliud est, quàm bona totius corporis habitudo et conveniens singularum

by commonplace men. And of their natures and tempers, which it is necessary for a physician to know, all but learned men are ignorant. And so this is the first reason; it is drawn from the subject of the medical art.

The other reason, which is taken from the end, is of course especially worthy of consideration but, it appears, utterly unknown to others. For this altogether forces the physician, in conveying medical art, to commence with making the human body and its parts clear. And he cannot do otherwise at all, if he is going to use resolutive order from a notion of the end. And indeed the end has to be known beforehand by a twofold prior knowledge, just as Aristotle teaches about the subject of a science in the first book of the *Posterior Analytics*:[51] As in a science the subject has to be known beforehand — both that it is, and what it is — so in an art the end ought to be known° beforehand — both that it is or could be made, and what it is. For unless it could be, the means for its generation would be sought in vain. And if we were ignorant of what it is, then we could in no way discover those means, [since] we would lack a beginning (*principium*) fitting for their discovery — which is just the definition of the end. For from this one thing [i.e., the definition of the end] we can be led to knowledge of the means or beginning-principles (*principia*). Aristotle teaches this in text no. 23 of the seventh [book] of the *Metaphysics*,[52] saying that the teaching of all arts is done from a notion of the end and that a notion of the end is nothing other than the definition of the end itself.

Now the end of the medical art is health, both conserving it if it is present and restoring it if it has lapsed. The physician, therefore, ought to know° before all else that health can be lost and restored — this is, of course, very well known to everyone — and then what health is and what it consists in. For if he is ignorant of this, how will he be able to track down remedies for preserving and restoring health? For health is nothing other than the good condition of the whole body and the appropriate temper of each of the

7

8

partium temperies. Constat enim aliam esse elementorum misturam in carne, aliam in nervo, aliam in osse; et aliam in sanguine, aliam in bile; proinde diversos esse in singulis partibus primarum qualitatum gradus diversasque temperaturas, in quarum conservatione cum quadam latitudine sanitas est constituta. Nam si omnes homogeneae partes propriam retineant naturalem temperaturam, heterogeneae, quae ex his constant, muneribus suis optimè fungentur, dummodò adsit etiam debita singularum partium figura et situs et magnitudo tum simpliciter tum ipsarum inter se comparatione. Hoc certè vel tale aliquid sanitas est, diligentem enim, et exquisitam ipsius definitionem constituere non est praesentis contemplationis. Satis nobis est in praesentia ostendisse quòd sanitas non potest benè cognosci, nisi omnes humani corporis partes, earumque naturae et temperies et officia et operationes cognoscantur.

9   Unde manifestum est tantum abesse ut ars medica tradita ab Avicenna servet ordinem compositivum, ut omnes putant, cùm ab elementis incipiat et ab humoribus aliisque corporis humani partibus, ut potius hac ipsa ratione dicatur scripta ordine resolutivo. Est enim propria et essentialis conditio ordinis resolutivi, ut ante omnia praecognoscatur tum finis, qui efficiendus est, tum subiectum, in quo est efficiendus. Quae quidem praecognitio in aliis artibus magis, in aliis minùs exquisita requiritur pro earum diversis conditionibus et naturis. Ars certè medica eam valdè exactam postulat, nam quid sit sanitas non satis pro ipsius conservatione vel recuperatione intelligere possumus, nisi causas et materiales et finales omnium partium humani corporis cognoscamus, finales quidem causae sunt singularum officia et operationes; materiales verò quatuor prima corpora, quae elementa vocantur, ex[38] quorum diversis commistionibus partes nostri corporis singulae diversas et proprias temperaturas adipiscuntur. Ad primam autem materiam

parts. For it is evident that some of the mixed elements are in flesh, some in nerve, some in bone, and some in blood, some in bile, and accordingly there are different degrees and different balances of primary qualities in each of the parts. Health is constituted in the conservation of these, with some latitude. For if all the homogenous parts retain their proper natural balance, the heterogeneous, which are composed out of these, will perform their jobs optimally, as long as the obligatory shape, arrangement, and magnitude of each of the parts, both absolutely and in comparison with each other, are also present. Certainly this or some such [thing] is health. To establish a careful and fine definition of it is not the present issue being contemplated. At present it is enough for us to have shown that health cannot be known well unless all parts of the human body, and their natures, tempers, functions, and activities, are known.

From this it is manifest that the medical art conveyed by Avi- 9 cenna maintains a compositive order so little—as everyone holds, since he starts from the elements and humors and other parts of the human body—that by this very reasoning it may be said to be written using resolutive order instead. For it is a proper and essential characteristic of resolutive order that, before all else, both the end that is to be effected and the subject in which it is to be effected has to be known beforehand. This prior knowledge, of course, is required to be finer in some arts, less so in others, according to their different characteristics and natures. Certainly medical art demands that it be highly exact. For we cannot understand what health is enough for its conservation and restoration, unless we know both the material and final causes of all parts of the human body: the final causes, of course, are the functions and activities of each, and the material are the four first bodies, what are called the elements, through different commixtures of which each of the parts of our body obtain different and proper balances. The physician does not, however, arrive at first matter and

et substantialem formam medicus non pervenit, quia harum cognitione non eget, propterea medicina cadit à perfectione scientiae speculativae, ut alio in loco demonstravimus.

10   Videtur autem duplici via medicus uti ad cognoscendas humani corporis partes, una quidem per sensum et per anatomen, qua sine causarum cognitione rem ita esse cognoscit. Altera verò per rationem, quam ex naturali philosophia desumit. Sic etenim Aristotelem quoque de animalium partibus disseruisse conspicimus, rem namque varietatis ac difficultatis plenam esse cognoscens operae pretium fore duxit si harum rerum historiam praemitteret, in qua sine causarum redditione id solum, quod experientia ipsa ostendere potuit, de animalium partibus doceret. Deinde verò in libris de Partibus animalium rationem reddit eorum omnium, quae simplici enarratione in libris de Historia exposuerat. Ea namque est ingenii nostri imbecillitas, ut non statim integram rei notitiam° capessere valeamus,³⁹ sed gradatim progrediamur et à cognitione quòd sit, quam confusam vocant, ad cognitionem cur sit, quae distincta dicitur, transeamus.

11   Galenus igitur et Avicenna ab ordinis resolutivi conditione non recesserunt, dum in arte medica constituenda à cognitione humani corporis et eius partium inchoandum esse censuerunt, hoc enim est exordiri à cognitione sanitatis, quae finis est.

12   Non possum autem hac in re Avicennae iudicium non magnopere probare, qui in hac partium humani corporis consideratione dicit credendum in multis esse philosopho naturali, neque esse exquisita ratione omnia investiganda, veluti numerum et naturas elementorum, de quibus medicus disputare minimè debet, sed ea sumere ut declarata fusius à philosopho naturali. Idcirco

substantial form, because he does not need knowledge of them. Because of this, medicine falls below the perfection of speculative science, as we demonstrated in another passage.[53]

And so it appears the physician uses a twofold way to know parts of the human body. One [way] is by means of sense and by means of anatomy, by which he knows something such as it is, without knowledge of causes. And the other is by means of reason, which [he] draws from natural philosophy. And indeed we see that Aristotle discussed the parts of animals this way also. For knowing that the issue is full of disparity and difficulty, he came to think it would be worth the work if he set out in advance the history of those things; in this [history], without a rendering of causes, he taught only those things about parts of animals that he could show by experience itself. And then in the books *On the Parts of Animals*, he renders the reason for all of them, which, in a simple rehearsing, he had laid out in the books *On the History [of Animals]*. For it is [because of] the feebleness of our wit that we cannot at once get hold of a fully integrated knowledge° of the thing; instead we progress by degrees and move on from the knowledge that it is, which they call confused, to knowledge why it is, which is called distinct.

Galen and Avicenna, therefore, did not deviate from what is characteristic of resolutive order when, in establishing medical art, they deemed that it has to start with knowledge of the human body and its parts. For this is to begin with the knowledge of health, which is the end.

On this issue, however, I cannot approve Avicenna's judgment too highly. In this consideration of parts of the human body he says in many things the natural philosopher has to be believed, and not everything has to be investigated with fine reasoning—such as the number and nature of the elements. About these the physician ought not to debate at all but should take them as clarified at greater length by the natural philosopher. Therefore he

10

11

12

Galenum haec accuratius tractantem, quàm medico par sit, censet
ut philosophum potius, quàm ut medicum in ea parte legendum
esse. Nec mirum, propria namque Graecorum fuit verbositas, ac
orationis prolixitas, quam pauci admodum inter eos vitare potuerunt, ast ii pauci naturae miracula tum priscis temporibus ha-
biti sunt, tum ad haec usque tempora habentur.

13   Illud quidem omnino credere et confiteri debemus, omnium
naturalium rerum, suarumque causarum cognitionem ad naturalem philosophiam pertinere, è qua hauriri dicitur quicquid in aliis
disciplinis de natura corporum naturalium dicitur. Quod videntur
medici omnes confiteri, dum primam hanc medicinae partem
φυσιολογικήν appellant, quasi non propriè medicinalem, sed è
naturali philosophia desumptam, licèt non amplius naturalis vel
scientialis ea tractatio dici possit; cùm enim philosophus naturalis
de elementis tractationem scientialem faciat, nempè non ut materia humani corporis sunt, ea enim vitiosa et sophistica esset consideratio in scientia speculativa, sed ut materia mistorum omnium,
necnon de partibus homogeneis et heterogeneis prout animalium
partes sunt, non prout hominis tantùm, ut propositiones sint per
se et universales et veram scientiam pariant. Medicus in solo humano corpore, non in aliorum animalium corporibus conservare et
efficere sanitatem volens, eam totam doctrinam humano corpori
accommodat, quae non quidem scientialis sed potius artificialis et
scopo finique medici idonea nuncupanda est.

14   Hinc fit, ut multa paucis verbis à medico sint perstringenda
tanquàm fusius declarata in philosophia naturali et à medici operatione remotiora, ut quae ad elementa pertinent; multa etiam fusius
tractanda, veluti propriae partium temperaturae et membrorum
singulorum functiones et munera, his enim maximè usurus est
medicus. Propterea quae de his naturalis philosophus docuit in

deems⁵⁴ that Galen, treating these things more precisely than is right for a physician, has to be read in this part as a philosopher rather than as a physician. And no wonder, for verbosity and prolixity of speech was proper to the Greeks; very few among them could avoid this. And those few were held to be small wonders of nature, both in ancient times as well as up to our own.

Of course, we ought altogether to believe and confess that 13 knowledge of all natural things and of their causes pertains to natural philosophy; whatever is said in other disciplines about the nature of natural bodies is said to be drawn from it. All physicians appear to confess this when they call the first part of medicine *physiologia*, as if not properly medicinal, but taken from natural philosophy, even granted that the treatment can no longer be called natural or scientific. For the natural philosopher treats the elements scientifically, that is, not as they are the matter of the human body—for in speculative science this consideration would be flawed and sophistic—but as the matter of all mixed [bodies] and also treats the homogeneous and heterogeneous parts in that they are the parts of animals and not only in that they are [the parts] of man, so that the premises may be *per se* and universal and bring forth true scientific knowledge. But the physician, wanting to conserve and effect health only in the human body and not in the bodies of other animals, applies that whole teaching to the human body; this is, of course, not scientific, but rather practical, and has to be called fit for the goal and end of the physician.

And so it happens that many things have to be touched upon 14 by the physician with few words, things clarified at greater length in natural philosophy and more remote from the practical activity of the physician, such as those that pertain to the elements, and many things have to be treated at greater length, such as the proper balance of the parts and the functions and jobs of each member, for the physician will use these especially. And so what the natural philosopher taught about these things in animals, the

animalibus, debet medicus repetere in homine, nixus semper fundamentis naturalis philosophiae, è qua haec omnia desumens, ea ab universali et scientiali consideratione ad particularem et artificialem transferre dicitur, quippe operandi gratia, non sciendi. Ideò rectè annotavit Averroes in secundo libro Colliget, capite primo, tractationes medicorum non esse scientiales, neque demonstrativas, quia propositiones, quibus utuntur, non sunt per se et universales, attribuunt enim res universales subiecto particulari.

15   Ex his satis puto esse demonstratum qualisnam sit prima illa medicae artis pars, quae physiologica dicitur, in ea enim et subiecti et finis praecognitio traditur, quae accipitur ex naturali philosophia, ideò nulla ibi medicus utitur propria demonstratione, sed ipsam rei veritatem absque ulla probatione proponit. Aut si maioris declarationis gratia demonstratione aliqua utatur, eam sumit à philosopho naturali, ut dictum est.

16   Si haec est praecognitio finis, facile est ostendere, quomodo ars medica tradatur ordine resolutivo, ab hac enim cognitione ad remedia ascendere, quae sunt principia conservantia vel recuperantia sanitatem, est finem resolvere in sua principia, à quibus operatio medici postea incipit. Hoc autem, quomodo sit, mox in ipsis medicorum traditionibus facilius declarabimus, ne[40] idem frustra repetere cogemur.

17   Ut autem à notioribus ordiamur, primo loco de Averroe loquemur, qui manifestissimè et pulcherrimè servavit ordinem resolutivum; secundo loco de Avicenna; postremo de Galeno, de quo tum propter ipsius dicta tum propter aliorum sententias maximè dubia res est, cùm non ubique videatur artem medicam hoc uno ordine tradidisse.

physician ought to apply in turn to man; always relying on the foundations of natural philosophy, drawing from it all the things that [he] is said to transfer from universal and scientific consideration to the particular and practical, for the sake, of course, of practicing, not of knowing scientifically. Because of this, Averroës correctly noted in the first chapter of the second book of *Colliget*,[55] that the physicians' treatments are not scientific and not demonstrative, because the premises that they use are not *per se* and universal, for they attribute universal things to a particular subject.

From all this I hold that it is sufficiently demonstrated what 15 sort of thing that first part of medical art — what is said to be physiology — is. For in it is conveyed prior knowledge of both the subject and the end; this [prior knowledge] is accepted from natural philosophy. And so the physician there uses no proper demonstration and instead sets out the very truth of the thing without any proof. Or if for the sake of greater clarification, he uses any demonstration, he takes it from the natural philosopher, as was said.

If this prior knowledge is the end, it is easy to show how medi- 16 cal art is conveyed using resolutive order. For to ascend from this knowledge to remedies, which are the beginning-principles conserving or restoring health, is to resolve the end into its beginning-principles, from which the practical activity of the physician afterward starts. Now how this is, we will clarify soon and quite easily in [considering] what the physicians convey, lest we be forced to repeat the same thing for no reason.

So that we may begin with the more known, we will speak in 17 the first place about Averroës, who maintained resolutive order most manifestly and beautifully; in the second place about Avicenna; and finally about Galen, about whom there is great doubt, both on account of what he himself said and on account of the positions of others, since he does not appear to have conveyed medical art using this one order everywhere.

: XII :

In quo declaratur ordo resolutivus ab Averroe
servatus in medicae artis traditione.

1   Averroes in suo libro, qui dicitur Colliget, non modò usus est ordine resolutivo exquisitissimo; sed ipsum quoque ordinem declaravit et ostendit se non temerè et ab ipsa rei natura coactum eum
ordinem servasse, sed scientem et optimè praeditum logicae disciplinae cognitione, quam revera alii medici non benè calluerunt, ut
ipse ibi significare videtur. Ideò in praefatione eius libri protestatur
Averroes eos, qui non sint in arte logica et in naturali scientia eruditi, non esse intellecturos librum illum et artificium, quo est
conscriptus.

2   In primo igitur capite primi libri Averroes medicam artem arti-
ficiosè dispositurus et non ignorans alium ordinem artibus, alium
scientiis convenire, ante omnia fundamentum hoc statuere voluit,
medicinam non esse scientiam, sed artem effectricem, et reprehen-
dit definitionem medicinae à Galeno traditam in principio artis
medicinalis, in qua scientiam accipit ut medicinae genus, cùm potius artem accipere debuerit, qua de re nos loquemur inferius,
quando Galeni libros considerabimus.

3   Hoc fundamento constituto statim proponit Averroes ordinem,
quo ars medica disponenda est, quem dicit omnibus artibus effec-
tricibus competere quatenus artes sunt, quare manifestè asserit
artibus omnibus nullum alium ordinem convenire, quàm resolutivum, quia ex natura artis hunc ordinem sumit, ut nos quoque superius fecimus, quod autem alicui naturale ac essentiale est, id inseparabile et immutabile esse debet. Ordinem igitur medicinae
talem statuit, quemadmodum et artium omnium effectricium,

: XII :

In which the resolutive order maintained by
Averroës in conveying medical art is made clear.

Averroës, in the book of his that is called *Colliget*,[56] not only used   1
the finest resolutive order, but also made the order itself clear and
showed that he had not maintained this order rashly and as someone forced by the very nature of the subject, but as someone
knowing scientifically and someone optimally endowed with
knowledge of the discipline of logic, a knowledge other physicians,
in truth, were not well versed in, as he appears to indicate there.
And so in the preface of the book, Averroës insists that those who
are not learned in logical art and in natural science will not understand that book and the skill with which it was written.

   In the first chapter of the first book,[57] therefore, Averroës,   2
about to dispose medical art skillfully and not ignorant that one
order is appropriate to arts and another to sciences, wanted before
all else to fix this foundation: medicine is not a science, but an effective art. And he censured the definition of medicine conveyed
by Galen in the beginning of *The Art of Medicine*, in which he accepted science as the genus of medicine, since he ought instead to
have accepted art; we will speak about this below when we consider Galen's books.

   With this foundation established, Averroës at once[58] sets out   3
the order in which medical art has to be disposed. He says it [i.e.,
the order] applies to all effective arts insofar as they are arts. And
so he manifestly asserts that no order other than resolutive is appropriate to all arts, because he takes this order from the nature of
art, as we too did above. What, however, is natural and essential to
something, ought to be inseparable and immutable. He therefore
fixed such an order for medicine, just as for all of the effective arts.

dicit eam in tres tractationes esse dividendam hoc ordine dispositas, ut in prima de subiecto agatur, in quo est operandum, nisi enim artifex aliquam sui subiecti cognitionem habeat, ut medicus humani corporis, nil efficere vel docere potest.

4   In secunda verò de fine tractetur, quem vult artifex in subiecto illo efficere.

5   In tertia demum inquirantur instrumenta, quibus finem illum in eo subiecto efficere possimus,[41] haec autem sunt ipsa principia, ex quibus seu per quae artifex operatur, in quibus doctrina artis desinit et à quibus operatio incipit. Quare absque dubio tribuit artibus omnibus ordinem resolutivum, qui à notione finis tendit ad investiganda principia pro illius finis consecutione.

6   Tria haec capita° ab Averroe posita possumus nos ad duo redigere, quorum alterum complectatur principia artis docentis, quae sunt principia cognitionis; alterum verò posterius contineat ea, quae locum habent conclusionum, quarum collectio est finis ipsius artis docentis. Principia cognitionis sunt cognitio subiecti et cognitio finis. Conclusiones verò, quae ex hac cognitione investigantur, sunt ipsa principia rei, à quibus operatio incipit, hic enim est ordo resolutivus à finis notione ad invenienda principia, ex quibus seu per quae finis ille à nobis effici possit.

7   Non est autem praetereundum id, quod in illo primo capite doctissimè considerat Averroes, dicit enim principia cognitionis non posse in arte sua demonstrari, sed vel esse per se nota vel sumi ut declarata in alia aliqua disciplina, quod evenit in medicinae principiis, nam partes subiecti sunt per se notae, id est per se

He says it has to be divided into three treatments, disposed in this order: In the first, the subject in which the practicing is to be is dealt with. For unless the practitioner has some knowledge of his subject, as the physician does of the human body, he can effect and teach nothing.

In the second is treated the end that the practitioner wants to 4 effect in that subject.

Lastly, in the third, the instruments by which we can effect that 5 end in the subject are asked about. And these are the beginning-principles themselves, from which or by means of which the practitioner works, [and] in which teaching of the art finishes and from which practical activity starts. And so, without doubt, he ascribes to all the arts resolutive order, which from a notion of the end aims to investigate the beginning-principles for gaining that end.

We can reduce these three headings° given by Averroës to two: 6 one of these encompasses the beginning-principles of the art as taught, which are the beginning-principles of knowledge, and the other then contains those things that have the place of conclusions, the gathering of which is the end of the art itself as taught. The beginning-principles of knowledge are knowledge of the subject and knowledge of the end. And the conclusions, which are investigated from this knowledge, are the very beginning-principles of the thing, from which [beginning-principles] the practical activity starts. For this is resolutive order, from a notion of the end to discovering the beginning-principles, from which or by means of which that end can be effected by us.

Now not to be passed over is that which Averroës, in a most 7 learned way, considered in the first chapter.[59] For he says that the beginning-principles of knowledge cannot be demonstrated in their own art, but are either known *per se* or are taken as they are made clear in some other discipline; this happens in the beginning-principles of medicine. For the parts of the subject are known *per*

cognosci possunt, sensu namque per humani corporis sectionem cognoscuntur, finem autem dari nemo ignorat, omnes enim sciunt humana corpora et[42] sana esse et aegra. Quid autem sit sanitas et quid aegritudo, videtur quidem omnibus notum esse confusa quadam cognitione per nominalem quandam descriptionem, sed non definitione perfecta, quae ex causis constat et quae necessaria est pro tuenda vel recuperanda medicis auxiliis sanitate. Hoc autem sumit medicus à philosopho naturali, qui dat ei et cognitionem quatuor elementorum et primarum qualitatum et temperierum, quae in mistis prodeunt ex elementorum commistione. Itaque et humorum naturas et membrorum eorumque officia sumit medicus à philosopho naturali, ideò in tota[43] illa parte omnis redditio causae et omnis demonstratio per causam materialem vel per finalem tradita sumitur ex naturali philosophia.

8   Priorem igitur partem ad principia cognitionis attinentem tractavit Averroes in tribus libris, primo, secundo ac tertio, sed magno cum artificio, prius enim cognitionem subiecti tradit in primo libro, deinde cognitionem finis in secundo ac tertio, quam distinctionem neque à Galeno neque ab Avicenna factam comperimus, etsi enim ambo ab hac tractatione auspicati sunt, attamen subiecti declarationem à declaratione finis non seiunxerunt, quod pulcherrimè facit Averroes, in primo enim libro tractat anatomen totius humani corporis et docet diligenter singulas ipsius partes sine ulla declaratione temperierum vel officiorum, solùm enim exponit ex quot partibus totum corpus et singulum membrum constet, sine ulla redditione causae. Corpus autem et partes eius sunt subiectum in quo medicus est effecturus sanitatem. Tota haec cognitio à sensibus sumitur per humani corporis sectionem et pertinet ad cognitionem subiecti, ut ibi asserit Averroes.

se, that is, they can be known *per se*, for by sense they are known by means of a dissection of the human body. Moreover no one is ignorant that there is an end. For everyone knows human bodies are both sound and sick. Now what health is and what sickness is, appear, of course, to be known to all with some confused knowledge by means of some nominal description but not by a perfect definition, which is composed out of causes and which is necessary for preserving or restoring health by medical aids. The physician, however, takes this from the natural philosopher, who gives him knowledge of the four elements, and of the primary qualities, and of the tempers[60] that issue in mixed [bodies] from the commixture of the elements. Thus, the physician takes from the natural philosopher the natures both of the humors and of the members and their functions. Because of this, every rendering of cause, and every demonstration by means of material or final cause, conveyed in that whole part, is taken from natural philosophy.

Averroës, therefore, treated the first part, pertinent to beginning- principles of knowledge, in three books, first, second, and third, but with great skill. For first he conveys knowledge of the subject in the first book, then knowledge of the end in the second and third; we find this distinction made neither by Galen nor by Avicenna. For although both commenced with this treatment, nonetheless they did not separate clarification of the subject from clarification of the end, something Averroës does most beautifully. For in the first book, he treats anatomy of the whole human body and carefully teaches each of its parts without any clarification of tempers or functions. He just lays out how many parts the whole body and each member is composed of, without any rendering of cause. Moreover, the body and its parts are the subject in which the physician will effect health. This whole knowledge is taken from senses by means of dissection of the human body and pertains to knowledge of the subject, as Averroës there asserts.[61]

8

9   In secundo libro tradit nobis Averroes cognitionem finis, nempè sanitatis, ideò in prooemio totius operis, dum facit divisionem librorum, dicit eum appellari librum sanitatis, et in principio illius secundi libri ponit definitionem sanitatis dicens sanitatem esse bonam temperaturam in singulis partibus corporis humani, cum qua potest edere suas naturales operationes et propriis muneribus fungi. Deinde in toto illo libro declarat bonam et naturalem temperaturam tum corporis totius tum singularum partium et earum propria munera et proprias operationes. Naturalem autem cuiusque partis temperaturam dicit ubique esse illius partis sanitatem, confundit enim nomen sanitatis cum nomine temperaturae naturalis, quae proculdubio idem significant, si idem est definitio, ac definitum, sanitatis enim definitionem hanc esse dixerat, sanitas est bona temperatura, per quam omnes partes propria possunt[44] officia exercere.

10   Acturus autem in eo libro de singularum partium temperaturis incipit in primo capite à quatuor elementis, quia temperatura nil aliud est, quàm forma complexionalis (ut ait Averroes) prodiens ex commistione quatuor elementorum, quare necessaria est cognitio elementorum et qualitatum primarum ad temperaturam totius corporis, ac singularum partium cognoscendam, proinde ad cognitionem sanitatis. Tractatio igitur de elementis est pars tractationis de sanitate et pertinet ad notionem finis.

11   Cum secundo libro coniungendus est tertius, qui est de aegritudine, nam is quoque pertinet ad notionem finis, ut Averroes ipse in primo capite primi libri asserit, eadem enim est cognitio contrariorum et qui novit sanitatem, aegritudinem quoque cognoscat necesse est, ut benè Averroes in primo capite illius tertii libri declarat, est enim aegritudo mala temperies laedens operationem ipsius membri, quae definitio colligitur, ut Averroes ibi ait, ex

In the second book, Averroës conveys to us knowledge of the 9
end, that is, of health. And so in the proem of the whole work,[62]
when he makes a division of the books, he says it is called the
book on health. And he places the definition of health in the beginning of that second book,[63] saying that health is a good balance
in each of the parts of the human body; with this it can bring
about its own natural activities and perform its proper jobs. Then
in the whole book, he clarifies the good and natural balance both
of the whole body and of each of the parts, and their proper jobs
and proper activities. Moreover, he says everywhere that the natural balance of any part is the health of that part, for he conflates
the name of health with the name of natural balance, which without doubt signify the same thing, if the definition and the defined
are the same. For he had said this is the definition of health:
health is a good balance by means of which all parts can exercise
their proper functions.

When he is about to deal in this book with the balances of each 10
of the parts, he starts in the first chapter from the four elements,
because balance is nothing other than a complexional form[64] (as
Averroës says) issuing from a commixture of the four elements.
And so knowledge of the elements and of the primary qualities is
necessary for knowing the balance of the whole body and of each
of the parts, and thus for knowledge of health. The treatment on
the elements, therefore, is part of the treatment on health and
pertains to a notion of the end.

The third book, which is about sickness, has to be conjoined 11
with the second, for it too pertains to a notion of the end, as Averroës himself asserts in the first chapter of the first book.[65] For
knowledge of contraries is the same, and whoever knows health
also necessarily knows sickness, as Averroës makes very clear in the
first chapter of that third book.[66] For sickness is a bad temper in-
juring the activity of the member itself; this definition is gathered,
as Averroës there says, from the definition of health. From this it

definitione sanitatis. Unde manifestum est aegritudinem in arte
medica habere locum finis; est enim finis medici sanitas tuenda vel
recuperanda, quem sensum aliis verbis eundem exprimimus, si dicamus finem esse aegritudinem vitandam vel repellendam.

12   In quarto libro agit Averroes de signis sanitatis et aegritudinis,
haec autem sunt accidentia consequentia bonam et malam partium
temperiem, quae diversa ratione à medico et à philosopho naturali
considerantur, philosophus enim naturalis quemadmodum sanitatem et morbum et ipsorum causas cognoscere vult tanquàm accidentia naturalia viventium corporum, ita etiam vult cognoscere
alia accidentia posteriora, quae ex illis prioribus prodeunt, medicus verò qui cognitionem accidentium naturalium pro fine non
habet, sed operationem, quae non est nisi in rebus particularibus,
ita his accidentibus utitur, ut ei ad operandum inserviunt. Operatio autem est in hoc et illo homine sanitatem tueri vel inducere seu
aegritudinem vitare vel repellere. Id autem praestare non potest,
nisi in singulis hominibus sanitatem et aegritudinem et aegritudinis speciem internoscat. Iis igitur accidentibus utitur tanquàm
signis sensilibus causas occultiores, nempè sanitatem et morbum
indicantibus,° accidentia enim tunc vocantur signa, quando causis
suis notiora sunt et per ea media ducimur in causarum cognitionem.

13   Pars haec de signis sub primam partem, quae de principiis cognitionis dicitur, reduci potest, pertinet enim ad tractationem de
sanitate et morbo et ad eorum in singulis hominibus cognitionem.
Quia postquàm universè sanitatis et aegritudinis naturam, cau-
sasque cognovimus, si universalem morbum curare et universalem
hominem sanare oporteret, nulla certè signorum cognitione opus
esset, iam enim certa morbi natura proponeretur, cui contraria remedia inquirenda et adhibenda essent.

14   Praeterea cùm particularis morbus et particularis homo sanan-
dus proponatur, si statim morbus et morbi genus internosceretur

is manifest that sickness has the place of an end in medical art, for the end of the physician is preserving or restoring health; we say this same sense with other words expressly if we say that the end is avoiding and repelling sickness.

In the fourth book,[67] Averroës deals with the signs of health and sickness. These, however, are an accident following on the good and bad temper of the parts, which are considered in different ways of reasoning by the physician and by the natural philosopher. For the natural philosopher wants to know health, disease, and their causes inasmuch as they are natural accidents of living bodies and so also wants to know other posterior accidents that issue from these prior [ones]. The physician, on the other hand, who does not have knowledge of natural accidents as an end but practical activity (which is only in particular things), thus uses these accidents as they serve him in practicing. And the activity is to preserve or induce health, or avoid or repel sickness, in this and that man. But he cannot really do this, unless he distinguishes health, sickness, and the species of sickness in each man. He therefore uses these accidents as sensible signs indicating° hidden causes, that is, health and disease. Accidents then are said to be signs when they are more known than their causes and we are led by means of the middles into knowledge of causes.

This part about signs can be brought back under the first part, which is said to be about the beginning-principles of knowledge, for it pertains to a treatment on health and disease and to knowledge of them in each man. Because after we have come to know the nature and causes of health and sickness universally, if universal disease needed to be cured and universal man healed, there would certainly be no need for knowledge of signs. For then the definite nature of disease would be set out, and remedies contrary to it would be sought and applied.

But when a particular disease and a particular man are set out to be healed, if the disease and the genus of the disease were

et ipse per se sensilis esset, similiter supervacanea esset tota haec de signis tractatio. At quia sanitas et morbus in hoc vel illo homine à medico non discernuntur sensu, ideò signis sensilibus opus est, quibus à sanitate morbus et morbi species dignoscatur.

15   Manifestum est igitur, quòd si nullus alius esset medicae artis scopus, quàm à notione finis invenire principia, eaque cognoscere, de signis agere non oporteret. Sed quia ultimus ac praecipuus scopus non est cognoscere, sed ex cognitione remediorum operari, necessaria fuit tractatio de signis propter artis medicae usum et operationem, non propter inventionem principiorum, ad hanc enim nihil confert.

16   Cùm autem ob dictam rationem debuit medicus agere de signis sanitatis et morbi, nullus fuit convenientior locus huius tractationis in arte medica, quàm statim post ipsam tractationem de sanitate et morbo, cùm ex ipsa sanitatis ac morbi natura signa haec tanquàm effectus à causis suis deriventur. Ars enim universalia praecepta docens à causis ad afferendam signorum rationem° progreditur instar philosophiae naturalis, ex qua haec omnia sumuntur. Contra verò ars operans à signis ad cognoscendam sanitatem et aegritudinem progreditur in hoc vel illo homine sanando.

17   Quatuor igitur priores libri totam partem ad principia cognitionis attinentem complectuntur, eamque ita perfectam et omnibus numeris absolutam et optimo ordine dispositam, ut nihil prorsus in ea desiderari posse videatur.

18   Tribus autem posterioribus libris altera pars continetur, in qua à proposito fine ad causas conservantes et restituentes sanitatem ascenditur, quarum inventione absolvitur ordo resolutivus. In his quoque ordinem servat Averroes convenientissimum, nam primo

at once distinguished and it were sensible *per se*, then, similarly, this whole treatment on signs would be superfluous. But because health and disease in this or that man are not discerned by the physician by sense, sensible signs are needed by which the disease and the species of the disease may be distinguished from health.

It is manifest, therefore, that if the goal of medical art were 15 nothing other than to discover beginning-principles from a notion of the end, and to know them, there would be no need to deal with signs. But because the ultimate and principal goal is not to know but to work from knowledge of remedies, a treatment on signs was necessary, for the sake of the use and practical activity of medical art and not for the sake of the discovery of beginning-principles; for it does not contribute to this.

Now since, for the reason said, the physician ought to deal with 16 signs of health and disease, no place for this treatment in medical art was more appropriate than right after the very treatment on health and disease, since these signs are derived from the very nature of health and disease, as effects from their causes. For an art teaching universal precepts progresses—just as natural philosophy, from which all these are taken, [does]—from causes to bringing forward an account° of signs. An art as practiced, on the other hand, progresses from signs to knowing health and sickness in this or that man being healed.

The first four books, therefore, cover the whole part perti- 17 nent to the beginning-principles of knowledge—and that [part] so perfect and on all counts so complete and disposed using such a very good order, that it appears nothing else at all could be desired in it.

In the three later books is contained the other part, in which 18 there is an ascending from the given end to the causes conserving and recovering health, by discovery of which resolutive order is completed. In these [books], too, Averroës maintains the most

loco loquitur in quinto libro de naturis et viribus tum ciborum tum medicamentorum et prius quidem universè, deinde speciatim, ut exempli gratia prius loquitur de natura medicamentorum indurantium et mollientium et aperientium et aliorum huiusmodi, postea transit ad nominandas singillatim medicamentorum species, quibus hae facultates insunt.

19 Sed medici priorem quidem partem appellandam censent de remediis, posteriorem verò de materia remediorum, ut aliàs notavimus; quam sententiam nos quidem non reprehendimus, attamen ipse Averroes nostra potius utitur appellatione, nam in eo quinto libro in calce capitis 30 et in principio 31 priorem partem vocat universalem tractationem de cibis ac de medicamentis, posteriorem verò dicit esse specialem. Quare processum illum à generibus ad species esse existimavit, ut revera est.

20 Sic etiam Aristoteles in secundo capite primi libri Posteriorum Analyticorum à definitione demonstrationis progreditur ad quasdam universales principiorum conditiones inveniendas et declarandas. Postea verò in quarto capite declarat quaenam sint illae propositiones, in quibus omnes conditiones reperiuntur.

21 Demum Averroes in duobus postremis libris docet modum operandi per illa principia, videlicet per cibos et per medicamenta, in sexto enim libro docet, quomodo sanitas conservetur, deinde in septimo, quomodo amissa recuperetur. Itaque convenientissimo ordine hanc de principiis medendi tractationem disposuit, prius enim in quinto libro eorum naturas simpliciter consideravit, deinde in sexto et septimo de eisdem locutus est cum relatione ad finem propositum, ad sanitatem et aegritudinem. Proposita namque

appropriate order. For in the first place he speaks in the fifth book[68] about the natures and powers both of nutriments and of medicaments, and first, of course, universally and then specifically. For the sake of example, he speaks first about the nature of hardening, softening, opening, and other medicaments of this type, and afterward passes to naming, one by one, the species of medicaments to which these faculties belong.

But the physicians, of course, deem that the earlier part should   19
be called *On Remedies*, and the latter *On the Matter of Remedies*, as we noted elsewhere; this is a position, of course, that we do not censure. Nevertheless Averroës himself uses our appellation instead. For in the fifth book, at the end of chapter 30 and the beginning of 31,[69] he calls the earlier part a universal treatment on nutriments and medicaments; but the latter he says is specific. And so he judged that the proceeding was from genera to species, as in truth it is.

So Aristotle too, in the second chapter of the first book of   20
the *Posterior Analytics*,[70] progresses from definition of demonstration to discovering and clarifying some universal characteristics of beginning-principles. And then afterward, in the fourth chapter,[71] he clarifies which are those premises in which all the characteristics are found.

Lastly Averroës, in the final two books,[72] teaches the way of   21
practicing by means of those beginning-principles, that is, by means of nutriments and medicaments. For in the sixth book, he teaches how health is conserved, and then in the seventh, how lost [health] is restored. He therefore disposed this treatment on the beginning-principles of healing using the most appropriate order. For first, in the fifth book, he considered their natures absolutely, and then, in the sixth and seventh, he spoke about them in relation to the given end, to health and sickness. Then, for the

naturali temperie docet remedia, quibus conservetur; et proposita qualibet intemperie et aegritudine docet remedia, quibus expellatur.

22   Alia quaedam notatu digna de artificio Averrois consideranda manent, quae mox opportuniore loco expendemus, nunc satis sit in ipsa Averrois tractatione docuisse conditiones ordinis resolutivi, quem in illo libro exactissimè servavit Averroes.

23   Possumus autem apud eum quinque illas[45] medicinae partes inspicere, φυσιολογικήν, παθολογικήν, σημειωτικήν, ὑγιεινήν, θεραπευτικήν, quas medici semper in ore habent, tamen cur in has partes ars medica dividatur et cur eo ordine disponendae sint, neminem video cognovisse, nisi solum Averroem, qui eas in suo libro ordinatissimè ac distinctissimè tractavit et ipsarum tum necessitatem tum convenientem dispositionem non aliunde sumpsit, quàm ex ipsa ordinis resolutivi natura, dum arti medicae tradendae ac disponendae is ordo applicatur. Quod cùm facilè ex iis, quae modò dicta sunt, colligere quisque possit, in eo declarando non morabimur.

: XIII :

In quo ostenditur quomodo Avicenna medicam
artem tradiderit ordine resolutivo.

1   Eundem ordinem, licèt cum parvo discrimine, apud Avicennam quisque observare potest; illud enim discrimen ipsam resolutivi ordinis naturam nulla ex parte labefactat.

2   Primus quidem Avicennae liber in quatuor partes divisus est, quarum duae priores priorem artis medicae partem, quae est de principiis cognitionis, complectitur, in iis enim et subiecti et finis

natural temper set out, he teaches the remedies by which it is conserved; and for any distemper and sickness set out, he teaches the
remedies by which it is expelled.

Some other things worthy of note regarding Averroës' skill re- 22
main to be considered; we will weigh these soon, in a more op-
portune place. For now let it be enough to have taught the characteristics of resolutive order in that treatment by Averroës', the
[order] Averroës maintained most exactly in that book.

We can now observe in it those five parts of medicine that phy- 23
sicians always have on their lips: *physiologia, pathologia, sēmeiōtikē,
hygieinos,* and *therapeutikē.* I do see, however, that no one has
known why medical art is divided into these parts and why they
have to be disposed using this order — except only Averroës, who
treated them in his book very orderly and distinctly, and took both
the necessity and appropriate disposition of them from nothing
other than the very nature of resolutive order, when this order is
applied to conveying and disposing medical art. Since anyone
could easily gather this from the things that were just said, we will
not tarry in clarifying it.

: XIII :

*In which it is shown how Avicenna conveyed
medical art using resolutive order.*

Anyone can observe the same order in Avicenna,[73] granted with a 1
little discriminating difference; this discriminating difference in no
part undermines the nature itself of the resolutive order.

Avicenna's first book is, of course, divided into four parts, of 2
which the first two cover the first part of medical art, which is
about the beginning-principles of knowledge; in them knowledge

et signorum cognitio traditur. Illae igitur duae quatuor prioribus Averrois libris proportione respondent. Duae verò posteriores posteriorem eius artis partem continent, quae est de auxiliis, in qua et de tuenda sanitate et de expellenda aegritudine praecepta traduntur, his igitur sextus ac septimus libri Averrois respondent. Sed tamen Averroes tractationem de auxiliis prius separatim fecit in quinto, quàm in sexto et septimo de valetudine tuenda vel recuperanda loqueretur. Avicenna verò cum his duabus partibus tractationem de auxiliis commiscuit.

3   Et quoniam ibi Avicenna de auxiliis non egerat, nisi generaliter et confusè, quemadmodum etiam Averroes in prima parte quinti libri; ideò statim secundum librum adiecit, in quo egit particulatim de omnibus simplicibus medicamentis, de quibus Averroes in secunda parte illius quinti libri tractavit.

4   Hic igitur Avicennae ordo est absque dubio resolutivus, quia est ab humani corporis et sanitatis et morbi praecognitione ad auxilia et ipsorum usum, tanquàm à notione finis ad medicae artis principia.

5   Avicenna verò quoniam in primo libro de anatome humani corporis et de aegritudinibus earumque curatione universaliter et confusè locutus erat, ideò in tertio et quarto libris de his omnibus particulatim ac distinctè disserere constituit, ut doctrinam confusam et imperfectam et ab usu aliquanto remotiorem perficeret et medicae operationi proximam redderet. In tertio quidem libro de singularum partium structura, propriisque morbis, in quarto verò de febribus atque harum omnium curatione diligentissimè locutus est, in quibus libris notare possumus ipsum ab ordinis resolutivi conditionibus nunquàm recessisse, praemittit[46] enim cuiusque membri anatomen; deinde eius temperiem naturalem, quae sanitas ipsius dicitur, declarat. Mox morbi naturam, tandem verò auxilia

of subject, end, and signs is conveyed. Those two, therefore, correspond in comparative relation to Averroës' first four books. The latter two contain the latter part of the art, which is about [medical] aids, and in which precepts about both preserving health and expelling sickness are conveyed. To these, therefore, Averroës' sixth and seventh books correspond. But Averroës, nevertheless, first made a treatment on [medical] aids separately, in the fifth, and then in the sixth and seventh, he spoke about preserving and restoring healthfulness. But Avicenna combined with these two parts a treatment on [medical] aids.

And since Avicenna had not dealt there with [medical] aids, 3 except generally and confusedly, as Averroës [did] in the first part of the fifth book, he therefore added a second book straightaway; in it he dealt, one by one, with all the simple medicaments, which Averroës treated in the second part of the fifth book.

Avicenna's order, therefore, is without doubt resolutive, because 4 it is from prior knowledge of the human body, of health, and of disease to [medical] aids and their use, as from a notion of the end to the beginning-principles of medical art.

But whereas Avicenna had spoken in the first book universally 5 and confusedly on anatomy of the human body and on sicknesses and their cure, he then decided in the third and fourth books to discuss all these things distinctly and particularly, so that he could perfect the confused and imperfect teaching, so remote from any use, and render it proximate to practical medical activity. In the third book, he first spoke most carefully about the structures of each of the parts and about their proper diseases, and then in the fourth about the fevers and cure for them all. In these books, we can note, he never deviated from the characteristics of resolutive order. For he sets out in advance the anatomy of each member and then makes clear its natural temper,[74] what is said to be its health. Next he teaches the nature of disease and finally the [medical] aids

et curationem docet. Quare in singulo morbo speciatim tractando procedit à notione finis, est enim finis ipse morbus expellendus et transit ad principia, ad remedia, per quae expellatur.

6   Demum in quinto libro loquitur de medicamentorum compositione.°

7   Possumus igitur inter Averroem et Avicennam discrimen hoc notare, quòd Averroes totam considerationem anatomes[47] humani corporis in uno, atque eodem primo libro absolvit, ut etiam totam de aegritudinibus tractationem in tertio libro fecit, totamque de sanitate, ac naturali temperie in secundo. Avicenna verò in prima et secunda primi libri partibus perfectam horum tractationem non explevit, sed eam confusam et universalem reliquit, postea verò in tertio et quarto libris accuratè, singillatim omnia explicavit.

8   Huius autem discriminis rationem eam esse puto, quòd Avicenna fusius ac prolixius, Averroes verò brevius ac strictius[48] et veluti per compendium medicam artem tradere constituit. Quoniam enim prolixè et summa cum diligentia scribere artem illam voluit Avicenna, ideò satis habuit in primo libro totam artem confusè nobis ante oculos ponere et totius artis universalem habitum in mente nostra generare, ne singulorum exquisita et particulari declaratione, proinde nimia sermonis prolixitate nos taedio ac fastidio afficeret antequàm ad optatum et expectatum finem nos perduceret, nempè ad auxiliorum cognitionem, quae tota ultimo loco tradenda erat, quod quidem Averroes propter brevitatem vereri minimè debuit. Hanc igitur ob causam Avicenna particularem et distinctam omnium tractationem ad tertium et quartum librum remittere voluit, ob[49] artis perfectionem et complementum.

9   Propterea dicere possumus in totius artis traditione Averroem semel usum esse ordine resolutivo, Avicennam verò pluries, semel

and cure. Whereby in treating [i.e., discussing] each disease specifically, he proceeds from a notion of the end—for the end is indeed the expelling of the disease—and passes to the beginning-principles, to the remedies, by means of which it is expelled.

Lastly, in the fifth book, he speaks about the combining° of 6
medicaments.

We can, therefore, note this discriminating difference between 7
Averroës and Avicenna: Averroës completed the whole consideration of anatomy of the human body in one and the same first book, as he made the whole treatment on sickness in the third book and the whole one about health and natural temper in the second. But Avicenna, in the first and second parts of the first book, did not provide a perfect treatment of them; he left it confused and universal. And then afterward, in the third and fourth books, he explicated all of them precisely, one by one.

I hold that the reason for this discriminating difference is that 8
Avicenna decided to convey medical art fully and prolixly, while Averroës [did so] only briefly and in a more compressed way and as it were by means of a compendium. And since Avicenna wanted to write the art prolixly and with utmost care, he was satisfied, in the first book, to put the whole art confusedly before our eyes and to generate in our mind the universal habit of the whole art, lest, with a fine and particular clarification of each thing, he affect us with tedium and distaste by the great prolixity of the discourse, before he could lead us on to the desired and expected end, that is, to knowledge of the [medical] aids, which was the whole of what was to be conveyed in the last place; for the sake of brevity, of course, Averroës ought not to have worried about this at all. Because of this, therefore, Avicenna wanted to leave a particular and distinct treatment of everything for the third and fourth books, for the perfection and completion of the art.

We can say, therefore, that in conveying the whole art, Averroës 9
used resolutive order once, but Avicenna [did so] many times—

quidem in primo libro, cum quo etiam secundus coniungatur; saepius autem in tertio et quarto, nimirum toties, quot sunt morbi, quorum curationem docet.

10 Praeterea videbatur in tractatione Avicennae quintum librum secundo statim adiungendum esse, ut sermonem de simplicibus medicamentis statim de ipsorum compositione° tractatio sequeretur, tamen facilioris doctrinae gratia Avicenna ad eum postremum locum librum illum reiecit, cùm ea medicamentorum compositio° non videatur benè intelligi potuisse, nisi prius singularum partium temperies distinctè cognitae fuissent. Possumus autem tum hac in re tum in aliis, quae de dispositione Avicennae diximus, illud inspicere, quod suprà, cùm de ordine generaliter loqueremur, annotavimus, saepe in disciplinis evenire ut propter nostram faciliorem cognitionem varietur aliqua ex parte is ordo, quem ipsa rerum tractandarum natura postulabat. Per hanc tamen variationem non stat quin scientiae omnes ordine compositivo scriptae dicantur, artes verò omnes resolutivo. Nam et in artis medicae traditione nullum alium ordinem ab Avicenna servatum fuisse ostendimus, quàm resolutivum.

11 Ex his manifestus factus est omnium medicorum error in declarando ordine ab Avicenna servato, quoniam enim ab elementis auspicatur et ab his transit ad humores, deinde ad alias corporis partes natura posteriores, proinde à simplicibus ad composita, putant omnes ordinem ab eo servari compositivum. Nullum est aliud ibi Avicennae consilium, quàm scientiam scribere naturalem et nobis cognitionem fabricae° humani corporis tradere. Nam si hic solus ipsius scopus fuisset, utique ab elementis inchoando usus esset ordine compositivo, eiusmodi tamen tractatio non amplius artis medicae pars fuisset, sed scientiae naturalis.

once in the first book, of course, with which the second is also conjoined, and more frequently in the third and the fourth, indeed as many times as there are diseases whose cure he teaches.

Moreover, it appeared that the fifth book in Avicenna's treatment should have been connected straight to the second, so that a discourse on combining° them would follow straightaway a treatment on the simple medicaments. But for the sake of easier teaching, Avicenna left that book for the final place, since it does not appear that combining° of the [simple] medicaments could have been well understood unless the tempers of each of the parts had first been known distinctly. Now both in this issue and in the other things we said about Avicenna's disposition, we can observe that which we noted above when we spoke about order generally, that in disciplines it often happens that on account of our easier knowledge, the order, which the very nature of the things being treated demands, is changed in some part. But saying that all sciences are written using compositive order and all arts using resolutive is not stopped by this change. For we showed that in conveying medical art, no order other than resolutive had been maintained by Avicenna.

From all this the error of all the physicians in clarifying the order maintained by Avicenna is made manifest. For since he commences with the elements and passes from these to the humors, and then to the other parts of the body, which are by nature posterior, and thus from simples to composites, they all hold that the order maintained by him is compositive. Avicenna's intent there is nothing other than to write natural science and convey to us knowledge of the construction° of the human body.[75] And if this had been his only goal, then by all means he would have used compositive order, starting with the elements, but then a treatment of this type would no longer be part of medical art; it would instead be part of natural science.

12   Cùm autem is non fuerit Avicennae scopus in ea parte, sed
scopus fuerit à cognitione humani corporis et partium eius et sanitatis et morborum omnium progredi ad invenienda remedia, per
quae à medico operante sanitas praesens conservari et amissa recuperari queat; ideò tractatio de elementis et de humoribus et de
partibus et de temperiebus partium humani corporis est tractatio
de subiecto et de fine artis medicae, quorum cognitio ad auxiliorum inventionem et ad operationem, quemadmodum diximus,
necessaria est. Est igitur ordo resolutivus, quia est à notione finis
ad principiorum inventionem.

13   Certè huiusce rei veritas clarissima est et miror quòd medici
omnes ex sola resolutivi ordinis consideratione eam non inspi-
ciant.° Hunc enim ordinem esse à notione finis et ad principia
procedere ipsi non inficiantur, at cùm principii nomen respectum
notet ad aliquid, principium enim est alicuius principium, cuiusnam rei principia sunt illa, ad quae invenienda progredimur ordine
resolutivo? Nonne principia finis propositi, id est, per quae illum
producere possimus?⁵⁰ Ergo à sanitate ad principia sanitatis progredimur, quando ab eius notione ad investiganda remedia transi-
mus. Causae namque et principia, per quae medicus efficit sanitatem, sunt ipsa remedia, non quatuor elementa. Nam illa principia
ab arte docente inveniuntur ordine resolutivo, à quibus incipit
operatio ordine compositivo, ut ait Aristoteles in contextu 23 septimi Metaphysicorum à nobis saepe memorato. Medicus autem
per remedia operatur, non per elementa ut humanum corpus
constituentia. Elementorum igitur consideratio non est tractatio
de principiis respectu finis ipsius artis medicae, sed potius est pars

Since, however, this was not Avicenna's goal in that part — the   12
goal was to progress from knowledge of the human body and of its
parts and of the health and diseases of them all to discovering
remedies by means of which health when present can, by a practicing physician, be conserved and, if lost, restored — the treatment
on the elements and the humors and the parts and the tempers of
the parts of the human body is a treatment on the subject and on
the end of medical art, knowledge of which is necessary for discovery of [medical] aids and for practical activity, just as we said. It is,
therefore, resolutive order, because it is from a notion of the end to
discovery of beginning-principles.

Now certainly the truth of this issue is very clear, and I wonder   13
at the fact that all the physicians do not see° it just from consideration of resolutive order. Now they do not deny that this order is
from a notion of the end and proceeds to beginning-principles.
But since the name, "beginning-principle (*principium*)," marks a relationship to something — for a beginning-principle is a beginning-
principle of something — what are they beginning-principles of, to
discovery of which we progress using resolutive order? Are they
not the beginning-principles of the given end, that is, [those] by
means of which we can produce that [end]? Therefore we progress
from health to the beginning-principles of health when we pass
from a notion of it to investigating the remedies. For the causes
and beginning-principles by means of which the physician effects
health are the remedies themselves, not the four elements. For
those beginning-principles are discovered from the art being
taught using resolutive order, and practical activity starts from
them using compositive order, as Aristotle says in text no. 23 of the
seventh [book] of the *Metaphysics*,[76] often referred to by us. The
physician, however, works by means of remedies, not by means of
the elements as they constitute the human body. A consideration
of the elements, therefore, is not a treatment on beginning-principles with regard to the very end of medical art, but is rather part

tractationis de ipso fine, quia sanitas cognosci non potest sine cognitione naturalis temperiei, imò haec est sanitas ipsa, neque naturalis temperies sine cognitione primarum qualitatum et elementorum.

14   Sed dubitare quispiam posset an verum sit illud, quod modò dicebamus, medicum operantem à remediis exordium sumere, videtur enim incipere à consideratione signorum et ab his ad naturam morbi cognoscendam procedere, deinde ultimo loco remedia adhibere.

15   Dubium hoc facilè tolletur, si in arte medica operationem à cognitione dignoscamus, nec unam cum altera confundamus. Est igitur sciendum° talem esse medicae artis naturam, ut cùm operationem pro fine habeat, duas cognitiones operationi praecedere necesse sit, unam universalem, quae in arte docente traditur et de qua nos saepe locuti sumus, alteram particularem, quae statim cuiusque morbi curationem praecedit. Cùm enim ad aliquem aegrum curandum medicus primùm accersitur, necesse est ut antequàm efficere aliquid aggrediatur, morbi propositi naturam, genusque cognoscat, quo modo enim curare sciet,° si non cognoscat? Morbi igitur naturam inquirit tum per causas praecedentes tum per signa consequentia utens methodo resolutiva. Inventa morbi natura, in qua tamen cognoscenda saepe decipitur seu potius nos medicorum errore decipimur, iterum utitur methodo resolutiva à morbo ad remedia, quae ipsum expellere apta sint,[51] progrediens, quem quidem processum medici indicationem nominare solent. Hactenus medicus nihil efficit, nihil operatur, sed in sola cognitione adipiscenda versatus est, ut sequenti operationi, quae finis ultimus est, debitam et necessariam cognitionem praemittat. Adhibitio igitur remediorum est quidem finis respectu praecedentis

of a treatment on the end itself, because health cannot be known without knowledge of natural temper—indeed this is health itself—nor natural temper without knowledge of the primary qualities and the elements.

But someone could doubt whether what we just said, that the 14 practicing physician takes [his] beginning from remedies, is true; for he appears to start from a consideration of signs, and to proceed from these to knowing the nature of the disease, and then in the last place to apply remedies.

This doubt will easily be gotten rid of, if in medical art we dis- 15 tinguish practical activity from knowledge, and not conflate one with the other. It has to be understood,° therefore, that the nature of medical art is such that, although it has practical activity for an end, two [kinds of] knowledge must precede the activity: one universal, which is conveyed in the art being taught and about which we have often spoken; the other particular, which immediately precedes the curing of any disease. For when a physician is first summoned to cure some sick person, it is necessary that, before he undertakes to effect anything, he know the nature and genus of the given disease. For how will he know° how to cure [it] if he does not know [its nature and genus]? He asks, therefore, about the nature of the disease, both by means of the preceding causes[77] and by means of the consequent signs using resolutive method. The nature of the disease having been discovered—though he is often deceived in knowing it or rather we are deceived by the error of the physicians—resolutive method is again used, progressing from the disease to the remedies that are able to expel it. The physicians, of course, normally call this procedure indication. Up until now the physician effects nothing, works nothing, and is concerned only with obtaining knowledge, so as to set out the obligatory and necessary knowledge before the subsequent practical activity that is the ultimate end. The application, therefore, of remedies is, of course, the end with regard to preceding knowledge

cognitionis, sed est tamen principium medicae effectionis, à quo ad sanitatem progressio est ordo compositivus artis operantis, medicus enim per remediorum admotionem efficit sanitatem. Unde patet in arte medica contrarium semper esse ordinem operationis ordini cognitionis, ubi enim cognitio desinit, inde incipit operatio, sive universalem sive particularem cognitionem spectemus, utraque enim in remediis desinit, à quibus initium operationis sumitur.

16   Debemus autem ex iis, quae dicta sunt, distinctionem hanc colligere, ac memoriae mandare, quae omne dubium solvit et aliorum errorem declarat, duo principiorum genera in medicina considerantur, elementa quidem sunt principia respectu humani corporis, remedia verò sunt principia respectu sanitatis, quae est medicinae finis, ab his igitur ratio sumitur ordinis resolutivi, qui dicitur esse ad principia respectu finis, non respectu subiecti, illa enim sunt principia operationis, quemadmodum diximus, quod etiam Galenus significavit in libro suo de arte Medicinali, ubi de causis salubribus et insalubribus sermonem faciens remedia ipsa causas salubres, id est sanitatis causas appellat. De elementis autem nihil dicit, quia horum consideratio in medicina non est penitus necessaria, cùm possint supponi ut declarata à philosopho naturali, et continetur in tractatione de principiis cognitionis medicae artis, ut antè declaravimus.

but is nevertheless the beginning-principle of medical effectuation, progression from which [beginning-principle] to health is the compositive order of the art as practiced; for the physician effects health by means of the application of remedies. From this it is patent that in medical art the order of activity is always contrary to the order of knowledge. For where knowledge finishes, there the activity starts, whether we look at either universal or particular knowledge, for either one finishes in the remedies from which the start of practical activity is taken.

Now from the things that have been said, we ought to gather 16 and commit to memory this distinction that does away with every doubt and makes the error of others clear. Two kinds of beginning-principles are considered in medicine: first, the elements are beginning-principles with regard to the human body, and second, remedies are beginning-principles with regard to health, which is the end of medicine. The reason for resolutive order is taken from the latter; it [i.e., this order] is said to be toward beginning-principles with regard to the end, not with regard to the subject. For those are the beginning-principles of practical activity, just as we said, and as Galen indicated in his book *The Art of Medicine*. When discoursing on healthy and unhealthy causes,[78] he calls the remedies themselves healthy causes, that is, the causes of health. He says nothing, however, about the elements, because a consideration of them is completely unnecessary in medicine, since they can be taken as having been made clear by the natural philosopher, and [such a consideration] is contained in the treatment on the beginning-principles of knowledge in medical art, as we made clear earlier.

: XIV :

In quo ostenditur, Galenum in artis medicae traditione
alium ordinem non servasse, quàm resolutivum.

1   Quod verò ad Galenum attinet, si vera sunt ea, quae de Averroe et
de Avicenna diximus, satis facile nobis erit ostendere eundem ordinem non modò in libro de Artis medicae constitutione, sed in Arte
quoque medicinali à Galeno fuisse servatum.

2   Prius quidem, ut à notioribus ad ignotiora progrediamur, consi-
derandus est ordo ab eo traditus in libro de Artis medicae constitutione, quem certum est esse resolutivum, in primo enim et ultimo et penultimo capitibus eius libri dicit medicinam esse ex
numero artium effectricium earumque omnium constitutionem
esse à notione finis, quam dicit priscorum philosophorum sententiam fuisse.

3   Cùm igitur ibi Galenus profiteatur se ordinem servare resolutivum, videamus qualisnam ille ordo sit, quo medicam artem constituit, atque disponit; et à quibusnam incipiat et in quibus desinat.
Incipit ab humani corporis consideratione, omniumque eius partium declaratione et inquit omnes corporis nostri partes ante omnia cognoscendas esse, necnon singularum naturas et operationes
et magnitudines et situs. Hoc igitur vult esse exordium à notione
finis, cùm enim in hac tractatione versetur usque ad caput undeci-
mum, in eo capite epilogum facit omnium, quae dixerat, unde
colligit finem medicinae esse sanitatem tuendam vel recuperandam
et subiungit ad sanitatem tuendam eiusque labes curandas necesse
esse omnes corporis partes earumque naturas et officia cognoscere,

: XIV :

In which it is shown that in conveying medical art
Galen maintained no order other than resolutive.

Now what is pertinent to Galen is this: if the things that we said   1
about Averroës and Avicenna are true, it will be easy enough for us
to show that the same order was maintained by Galen not only in
the book *On the Constitution of the Medical Art* but also in *The Art of
Medicine.*

So that we may progress from [things] more known to [things]   2
more unknown, first to be considered, of course, is the order conveyed by him in the book *On the Constitution of the Medical Art,*
which certainly is resolutive. For in the first, last, and penultimate
chapters of the book[79] he says that medicine is among the number
of effective arts and that the constitution of all of them is from a
notion of the end; he says this was the position of the ancient
philosophers.

Though Galen, therefore, claims there that he himself main-   3
tains resolutive order, we can see of what sort is that order in
which he constitutes and disposes medical art, and from what
[things] he starts, and in what [things] he ends. He starts from
consideration of the human body and a clarification of all its parts
and says that all the parts of our body as well as the natures, activities, magnitudes, and arrangements of each have to be known
before all else. He wants, therefore, this beginning to be from a
notion of the end. For while he is concerned [with this] in this
tract up to the eleventh chapter, in that chapter he makes a summary of everything he had said,[80] from which he gathers that the
end of medicine is to preserve or restore health and adds that
for preserving health and curing its lapses it is necessary to know
all the parts of the body and their natures and functions and that

idque esse exordium à notione finis. In capite autem decimo praecedente declaraverat ex illis, quae antè dixerat, quaenam sit bona
corporis constitutio, quae sanitas dicitur et quae mala, quae dicitur
aegritudo seu insalubritas. Prius autem in capite octavo locutus
erat de elementis. In principio igitur artis medicae asserit Galenus
de elementis agendum esse, nec propterea putat talem ordinem
esse compositivum, sed resolutivum à notione finis, quia tractationem de elementis vult esse partem tractationis de sanitate, quae
consistit in bona primarum qualitatum commistione et temperie.

4   In capite autem duodecimo volens Galenus à⁵² sanitate iam declarata progredi ad inventionem principiorum, per quae sanitas
conservetur et amissa recuperetur, dicit in principio illius capitis
quòd curandi ratio à sanorum et aegrorum corporum affectione
auspicatur et transit ad investiganda remedia singulis morbis contraria, quae sunt principia operationis.

5   Has igitur duas partes in illa artis medicae dispositione conspicimus, unam, in qua de principiis cognitionis agitur, nempè de
cognitione subiecti et finis, alteram, in qua fit transitus ad principia rei, ad remedia, quae sunt causae sanitatis, hic enim revera est
ordo resolutivus.

6   In eo quoque libro, quem scripsit Galenus de ordine librorum
suorum, eandem sententiam protulit, inquit enim, ante alios esse
legendos Anatomicos libros. Idem significavit in libro illo, qui inscribitur, Quòd optimus medicus cognitione philosophiae praeditus esse debeat.

7   Haec cùm ita sese habeant, negare non possumus artem quoque
medicinalem eodem resolutivo ordine esse dispositam, in eo
namque toto libro de salubribus et insalubribus agitur, primùm

this is a beginning from a notion of the end. And in the preceding chapter, the tenth,[81] he had made clear, from those things that he had said earlier, what a good constitution of the body is — what is said to be health — and what a bad one is — what is said to be sickness or unhealthiness. But earlier, in the eight chapter,[82] he had spoken about the elements. Galen asserts, therefore, that the elements have to be dealt with in the beginning of medical art, and does not hold that because of this such an order is compositive; rather [he holds that it is] resolutive, from a notion of the end, because he wants a treatment on the elements to be part of a treatment on health, and that consists in a good commixture and temper of primary qualities.

Then in the twelfth chapter,[83] wanting to progress from health,   4
now clarified, to discovery of beginning-principles, by means of which health is conserved and, if lost, restored, Galen says, in the beginning of that chapter, that the rational way of curing commences with the affection of sound and sick bodies and passes to investigating the remedies contrary to each disease; these are the beginning-principles of practical activity.

In the disposition of medical art, therefore, we see these two   5
parts: one which deals with beginning-principles of knowledge, that is, with knowledge of the subject and the end; and another in which passage is made to the beginning-principles of the thing to the remedies that are the causes of health. And this in truth is resolutive order.

Also, in the book that Galen wrote about the order of his   6
books,[84] he advanced the same position. For he says that the anatomical books are to be read before the others. He indicated the same in the book that is titled, *That the Best Physician Ought to be Endowed with Knowledge of Philosophy*.[85]

We cannot deny these things, since they are as they are: *The Art*   7
of Medicine too is disposed using the same resolutive order, for in the whole book about the healthy and the unhealthy,[86] bodies, of

quidem de corporibus, deinde de signis, demum de causis. Atqui salubritas et insalubritas corporum est artis medicae finis, causae verò salubres remedia ipsa sunt, quae sanitatis principia esse diximus, in quibus desinit ordo resolutivus. Hic itaque est ille ipse ordo, quem diximus Averroem in suo libro servasse, primo enim loco de humano corpore ac de ipsius salubritate, atque insalubritate disseruit, ultimo loco de remediis, proinde de causis sanitatis tuendae et recuperandae, medio autem loco de signis tum sanitatis tum morbi. Cur igitur apud Galenum hic ordo non est resolutivus, si ille idem et apud Averroem et apud Galenum ipsum in libro de Artis medicae constitutione est ordo resolutivus? Ductus certè fuit Galenus ad hunc ordinem servandum ab ipsa artis natura, quae alio ordine se tradi non patitur, tamen insciens, ut videtur. Cùm enim proposuisset tradendam artem medicinalem ordine definitivo, quem à resolutivo diversum esse dixerat, tamen ab ipsa rei natura coactus incidit in ordinem resolutivum. Unde patet error multorum, qui putant ordinem illius libri reduci ad compositivum, quantum enim hi à veritate absint, satis ex iis, quae hactenus diximus, manifestum est. Sed Galeno quoque adversari videntur, qui definitivum ordinem ab aliis duobus distinxit, quare si putavit eum librum à se traditum esse ordine definitivo, non potuit existimare eundem ordinem esse compositivum.

8   Ipsum autem Galenum, si eum librum scripsit, quomodo defendere aut excusare possimus[53] equidem non video, eum enim ordinem servat, quem in libro de Artis medicae constitutione servandum esse dixit in artibus omnibus, qui est ordo resolutivus, ipse tamen alium ordinem esse putat, quem vocat definitivum.

9   Praeterea non tres dantur ordines, sed duo tantùm, hic enim ordo definitivus figmentum est, ut superius demonstravimus, quia definitio illa instar prooemii est, ordo autem in ipsis partibus

course, are dealt with first, then signs, and lastly causes. But healthiness and unhealthiness in bodies is the end of medical art, and the healthy causes are the very remedies that we said are the beginning-principles of health and in which resolutive order finishes. And this is the very order that we said Averroës maintained in his book. For in the first passage he discussed the human body and its healthiness and unhealthiness, and in the last passage the remedies and thus the causes for preserving and restoring health, and in the middle passage the signs both of health and of disease. Why, therefore, in Galen is this order not resolutive, if the same thing both in Averroës and in Galen himself in the book *On the Constitution of the Medical Art* is resolutive order? Certainly, Galen was led, however unawares, as it appears, to maintain this order by the very nature of the art, which does not allow itself to be conveyed using another order. For although he had set out to convey the art of medicine using definitive order, which he had said was different from resolutive,[87] [he] nevertheless fell into resolutive order, forced by the very nature of the thing. From this is patent the error of many, who hold that the order of that book is reduced to compositive. And how far they are from the truth is manifest enough from the things that we have said up until now. But they also appear opposed to Galen, who distinguished definitive order from the other two. For if he held that the book was conveyed by him using definitive order, he could not have judged that same order to be compositive.

Now, for my part, I cannot see how we can defend or excuse 8 Galen himself, if he wrote this book. For he maintains the order that in the book *On the Constitution of the Medical Art* he said has to be maintained in every art, which is resolutive order. But then he himself holds that it is another order, which he calls definitive.[88]

Moreover, there are not three orders, but only two. For this 9 definitive order is a figment, as we demonstrated above, because that definition is like a proem, and order is observed in the parts

tractationis attenditur, qua integra manente potest prooemium auferri.[54]

10   Sed neque reprehensione vacat ipsa per se Galeni definitio, cuius inventor licèt non ipse, sed Herophilus fuerit, ut alio in loco Galenus confitetur, attamen definitione vitiosa uti non debuit, praesertim dum eam ut fundamentum totius medicae artis construendae proponit.

11   Mihi quidem levis et parvi momenti esse videtur adversus hanc definitionem obiectio illa quòd medicinam inter scientias collocet, cùm non scientia sit, sed ars. Ad hanc enim satis ipse Galenus respondere videtur, dum ait se nomen scientiae non propriè, sed lato vocabulo accipere pro quacumque cognitione, haud enim falsò dicitur medicinam cognitionem quandam esse.

12   Sed illud in ea definitione vitium magnum inest, quòd rei definitae naturam et essentiam non exprimit, ars enim omnis in cognitione quidem[55] alicuius rei versatur, sed tamen ipsius natura non in cognitione, sed in operatione, ad quam cognitio tota dirigitur, consistit, de ipsa tamen operatione in ea definitione nihil dicitur. Propterea longè melior est definitio tradita ab Avicenna, qui dicit quidem medicinam esse cognitionem humani corporis prout sanitati vel morbo subiicitur, quae verba ad artis cognitionem pertinent, sed statim praecipuam definitionis partem subiungit, id est causam finalem, cuius gratia cognitio quaeritur, dum dicit, ut sanitas praesens conservetur et amissa recuperetur, quibus postremis verbis anima (ut ita dicam) medicinae continetur et sine illis natura ipsius non exprimitur. Patet autem definitionem traditam à Galeno esse priorem partem definitionis Avicennae dimissa parte posteriore, quae precipua est et omitti non debuit. Nec satis est si

themselves of the treatment, from which, [while] remaining intact, the proem can be removed.

But Galen's definition[89] is not itself *per se* free from censure, 10 even granted that its discoverer was Herophilus and not [Galen] himself, as Galen confesses in another passage.[90] Even so, he ought not to have used a flawed definition, especially when he set it out as the foundation of the whole medical art being constructed.

Now the objection to this definition, that it includes medicine 11 among the sciences although it is not a science but an art, appears to me, of course, to be of light and little moment. And Galen himself appears to respond enough to this when he says that he is [here] accepting the name of science not properly,[91] but as a broad term for any knowledge whatever, for it is not at all falsely said that medicine is some sort of knowledge.

But there is a great flaw in this definition, since it does not ex- 12 pressly state the nature and essence of the thing defined. Every art is, of course, concerned with knowledge of something, but its nature consists, nonetheless, not in knowledge but in the practical activity to which that whole knowledge is directed. But nothing is said in that definition about this activity itself. And so the definition conveyed by Avicenna is better by far. He says, of course, that medicine is knowledge of the human body in that it is subject to health or disease; these words pertain to the knowledge of the art, but he at once adds the principal part of the definition, that is, the final cause for the sake of which the knowledge is inquired after, when he says, "so that health when present may be conserved and when lost may be restored."[92] In these final words, the soul (so to speak) of medicine is contained, and without them its nature is not expressly stated. But now it is patent that the definition conveyed by Galen is the first part of Avicenna's definition with the latter part, which is the most important and ought not to have been omitted, put aside. It is not enough if someone says this is

quis dicat eam subintelligi, praecipua enim definitionis pars expressè proferenda est, non tantùm subaudienda.

13   Et multò melius faceret qui dimissa parte priore solam posteriorem sumeret dicens medicinam esse artem conservandi sanitatem sanis corporibus et eandem aegrotantibus restituendi.

14   Idcirco optima est definitio tradita ab Averroe in primo capite primi libri, Medicina est ars operatrix exiens à principiis veris, in qua quaeritur conservatio sanitatis corporis humani et remotio suae aegritudinis.

15   Definitio autem Galeni, si benè consideretur, ad philosophum naturalem potius, quàm ad medicum pertinet, philosophus enim naturalis res omnes naturales cognoscere vult earumque accidentia omnia, etiam sanitatem et morbum et horum signa et causas, ut antè dicebamus, ideò Aristoteles libellum scripsit de sanitate et morbo, cuius parvum fragmentum tantummodò habemus, reliqua desiderantur, quem librum non debemus medicum appellare, sed naturalem, quia cognitio cuiuscumque rei naturalis, quaecumque illa sit, ad naturalem philosophum pertinet, dum nulla nostra operatio cognitionem illam consecutura est. Sed cognitio alicuius rei naturalis, quae ad aliquam nostram operationem dirigatur, non amplius ad naturalem philosophiam, sed ad artem aliquam pertinet. Cuiusmodi discrimen inter scientiam naturalem et artem medicam in cognitione sanitatis et morbi declarat doctè Averroes in primo capite primi libri de Arte medica. Hanc autem differentiam facilè inspiceremus in definitionibus scientiae naturalis et cuiusque artis, nam in definitione artis exprimitur operatio, in definitione autem scientiae naturalis sola cognitio ponitur. Ideò si quis acciperet illam naturalis scientiae partem, quam scripsit Aristoteles de sanitate et morbo et eius definitionem assignare vellet, nulla certè

understood within it, for the principal part of a definition has to be advanced expressly and not just to be heard within it.

And someone would do much better by putting the first part 13 aside and taking only the latter, saying that medicine is the art of conserving health in healthy bodies and recovering the same in sickened ones.

Therefore the definition conveyed by Averroës in the first chap- 14 ter of the first book is optimal: "Medicine is the practical art proceeding out from true beginning-principles, in which the conservation of health in a human body and the removal of its sickness are inquired after."[93]

Galen's definition, on the other hand, if it is well considered, 15 pertains to the natural philosopher rather than to the physician. For the natural philosopher wants to know all natural things and all their accidents, including health, disease, and their signs and causes, as we were saying earlier. Aristotle, accordingly, wrote a small book on health and disease, of which we have only a little fragment; the rest is wanting.[94] We ought to call this book natural, not medical, because knowledge of any natural thing whatever, whatever it may be, pertains to the natural philosopher when no practical activity of ours is going to follow that knowledge. But knowledge of any natural thing, directed toward some practical activity of ours, no longer pertains to natural philosophy but to some art. In a scholarly way, in the first chapter of the first book in *On the Medical Art*,[95] Averroës makes clear a discriminating difference of this type between natural science and medical art in the knowledge of health and disease. And we were easily able to observe this differentia in the definitions of natural science and of each art. For practical activity is expressly stated in the definition of an art, while only knowledge is placed in the definition of natural science. And so if anyone were to accept that part of natural science that Aristotle wrote on health and disease and wanted to assign a definition to it, certainly no definition would be brought

alia definitio afferenda esset, quàm illa, quae à Galeno medicinae tribuitur, quia pars[56] illa scientiae naturalis est scientia salubrium et insalubrium et neutrorum, tum corporum tum signorum tum causarum. Quare non rectè aptatur medicinae haec definitio, sed omnino oportuit vel operationem vel saltem nomen artis in ea exprimere, siquidem artis nominatione operatio significatur, est enim ars habitus recta cum ratione effectivus.

16   Una remanet Galeni excusatio, quae quantum roboris habeat alii considerent, dicere possumus Galenum consultò et data opera artem operantem omisisse, cùm enim sibi proposuisset compendiariam tractationem facere artis medicinalis, solam artem docentem voluit ea definitione complecti dimissa brevitatis gratia arte operante, cùm clarum atque compertum omnibus sit, finem illius artis esse sanitatem servare vel recuperare, eruditis saltem, quibus potissimùm illud compendium scribitur.

17   Hanc fuisse Galeni sententiam ipse videtur in eo capite significare, cùm enim distinxisset salubre et insalubre in corpora et signa et causas, subiungit nomine salubris et insalubris causas praecipuè significari, deinde verò corpora et signa, quod quidem dictum de arte tantùm docente verum est, cuius scopus ac finis est per resolutionem invenire causas servantes et recuperantes sanitatem, finis enim semper dicitur principem locum obtinere. At de tota arte falsum esset, quandoquidem praecipua[57] est corporum salubritas, quae est ultimus medicinae finis et cuius gratia salubrium et insalubrium causarum cognitio quaeritur. Quod quidem Galenus in eodem capite confitetur, dicit enim in operatione primum locum corporibus tribui, quorum servanda vel recuperanda sanitas est; in cognitione autem primus locus tribuitur causis, hae namque sunt totius medicae doctrinae scopus et finis.

forward other than that which is ascribed to medicine by Galen, because that part of natural science is scientific knowledge of the healthy and of the unhealthy and of neither, — of bodies, of signs, and of causes. This definition, therefore, is not correctly applied to medicine. Instead, [he] altogether should have expressly stated in it either a practical activity or at least the name of an art, since practical activity is indicated by naming an art; for an art is a practical habit, made effective by right reason.

One excuse remains for Galen. How much weight it has, others   16
may consider. We can say that Galen intentionally and deliberately omitted the art as practiced. For since he had set out to make a compendious tract, *The Art of Medicine*, he wanted to have encompassed in the definition only the art as taught, putting the art as practiced aside for the sake of brevity, since to everyone it is clear and assuredly known that the end of that art is to maintain or restore health — or at least [it is] to the learned, for whom the compendium is mostly written.

That this was Galen's position he himself appears to indicate in   17
the chapter. For after he had distinguished the healthy and the unhealthy in bodies and in signs and in causes, he adds that by the names "the healthy" and "the unhealthy" causes are principally signified, but then bodies and signs.[96] But this dictum is, of course, true only of the art as taught, the goal and end of which is to discover by means of resolution the causes preserving and restoring health. For the end is always said to occupy the foremost place. But this would be false for the whole art, since there what is principal is the healthiness of bodies, which is the ultimate end of medicine and that for the sake of which knowledge of healthy and unhealthy causes is inquired after. Galen, of course, confesses this in the same chapter. For he says that in practical activity the first place is ascribed to the bodies whose health is to be maintained or restored; but in knowledge first place is ascribed to causes, for these are the goal and end of the whole medical teaching.

: XV :

In quo ex iis, quae dicta sunt, ordinum
numerus colligitur et Aristotelis atque
Platonis authoritate comprobatur.

1   Haec in arte medica et eius partibus diligentissimè considerare
volumus; ut in ea naturam resolutivi ordinis plenè declararemus.
Quae autem de hac arte diximus, aliis omnibus artibus aptanda
sunt. Quamvis enim in aliis exquisitam illam partium distinctionem non inspiciamus, tamen satis est, si in iis duas illas partes
consideremus, unam quidem[58] priorem, in qua de principiis cognitionis agitur, id est de fine, eiusque notione, necnon etiam de subiecto, si opus fuerit; alteram verò de principiis rei, nempè de illis, à
quibus operationem auspicando possimus finem illum consequi vel
producere.

2   Manifestum igitur est, nullum dari alium ordinem, quo aliqua
disciplina tradi possit, quam compositivum et resolutivum, ut ex
duplici disciplinarum natura demonstravimus, in contemplativis
enim, quae solam cognitionem pro fine habent, necesse est ordinem servare compositivum. In aliis autem omnibus, quoniam cognitionem ad finem aliquem à nobis producendum dirigunt, necessarium omnino est à notione finis ad principia progredi, qui est
ordo resolutivus. Itaque si tertium disciplinarum genus praeter ea
duo non datur, sequitur non dari praeter duos illos alium ordinem,
quo aliqua disciplina scribi vel tradi possit.

3   Sententiam hanc Aristoteles sectatus est, qui nullum unquàm
alium ordinem servasse comperitur, quàm compositivum et resolutivum; et nullius mentionem usquàm fecit, nisi horum duorum, nam

: XV :

*In which the number of orders is gathered from
the things that have been said and is confirmed
by the authority of Aristotle and Plato.*

We want to very carefully consider these things in medical art and   1
its parts, so that we may fully clarify the nature of resolutive order
in it. And what we said about this art has to be applied to all other
arts. For even though we may not observe that fine distinction of
parts in the others, it is nevertheless enough if we consider these
two parts in them: one, the first, in which the beginning-principles
of knowledge are dealt with, that is, the end and its notion, and
also the subject too, if it is needed; the other, [in which] the
beginning-principles of the thing [are dealt with], that is, those
things from which, by commencing practical activity, we can gain
or produce that end.

It is manifest, therefore, that there is no order in which any   2
discipline can be conveyed other than compositive and resolutive,
as we demonstrated from the twofold nature of disciplines. For in
contemplative ones, which have knowledge alone for an end, it is
necessary to maintain compositive order. But in all others, since
they direct knowledge to some end being produced by us, it is
altogether necessary to progress from a notion of the end to
beginning-principles; this is resolutive order. Thus, if there is no
third kind of discipline besides the two, it follows that there is no
order other than those two by which any discipline can be written
or conveyed.

Aristotle followed this position. It is found that he never main-   3
tained any order other than compositive and resolutive, and he
made mention of no other anywhere, except of these two. In the

in prooemio primi libri Physicorum et primi Mereorologicorum significat ordinem scientiae naturalis compositivum. In contextu autem 23 septimi Metaphysicorum et sexto libro de Moribus capite quinto necnon capite octavo septimi libri loquitur de ordine resolutivo, qui artibus et morali disciplinae conveniens est. De ordine autem definitivo vel de aliquo alio ne verbum quidem fecit unquàm Aristoteles.

4   Sed apertissimè sententiam suam profert in quarto capite primi libri de Moribus et significat eiusdem sententiae fuisse Platonem, inquit enim, 'ne lateat autem nos, differre eos sermones, qui à principiis sunt et eos, qui ad principia, rectè namque etiam Plato de hac re dubitabat et quaerebat utrum à principiis an ad principia via tenenda sit, quemadmodum in stadio ab Athlothetis ad metam, an è contrario.' Duos igitur Plato et Aristoteles ordines posuerunt, compositivum, quem vocaverunt à principiis et resolutivum, quem ad principia tendere dixerunt. Quòd si alium praeter hos duos ordinem dari existimassent, non rectè dubitassent, neque recta ipsorum quaestio fuisset, quia peccassent ob insufficientem ordinum enumerationem.

5   Res quoque ipsa declarat, alium praeter hos non dari doctrinae ordinem, ubi enim multa ordinatè tractanda sunt, quorum alia sunt aliorum principia, nullus alius videtur conveniens ordo servari posse, quàm vel à principiis ad ultima vel ab ultimo ad principia. Quod etiam illa Aristotelis comparatio significare videtur, in stadio enim si quis velit totum spatium ordinatè pertransire, non potest nisi vel à munerariis ordiendo progredi ad metam, vel contra à meta ad munerarios.

proem of the first book of the *Physics*[97] and [in that] of the first [book] of the *Meteorology*,[98] he indicates that the order of natural science [is] compositive. In text no. 23 of the seventh [book] of the *Metaphysics*,[99] and in the fifth chapter of the sixth book of the *Ethics*, and also in the eighth chapter of the seventh book,[100] he speaks about resolutive order, which is appropriate for arts and moral discipline. But about definitive order or any other, Aristotle never offered even a word.

He advances his own position very plainly in the fourth chapter   4
of the first book of the *Ethics* and indicates that Plato was of the same position. For he says, "Let this not escape us: Discourses that are from beginning-principles and those that are to beginning-principles differ. And Plato, too, correctly had doubts about this issue and inquired whether the way that has to be followed is from beginning-principles or to beginning-principles, just as in the stadium, [one must decide whether to proceed] from the judges to the goal cone, or vice versa."[101] Plato and Aristotle, therefore, posited two orders—compositive, which they called from beginning-principles, and resolutive, which they said aims toward beginning-principles. Now if they had judged that there was an order besides these two, they would not correctly have had doubts, nor would their question have been correct, because they would have gone astray on account of an insufficient enumeration of orders.

Moreover, the issue itself makes clear that there is not another   5
order of teaching besides these. For where many things have to be treated in order, of which some are the beginning-principles of others, it appears that no appropriate order can be maintained other than either from beginning-principles to things ultimate or from what is ultimate to beginning-principles. Even that comparison by Aristotle appears to indicate this, for in the stadium, if someone wants to pass across the whole field in order, he cannot do so unless either by beginning from the judges he progresses to the goal cone, or vice versa, from the goal cone to the judges.

6   Advertendum autem, ne repugnantiam in dictis Aristotelis esse suspicemur, qui eo in loco ordinem resolutivum dicit esse ad principia, sed in quinto capite sexti libri et in octavo septimi contrarium asserere videtur dicens in actionibus finem esse principium, sicut in mathematicis suppositiones. Ergo ordo resolutivus, cùm à fine exordiatur, à principiis est, si finis est principium.

7   At non sibi adversatur Aristoteles, quando enim ordinem resolutivum ad principia esse dicit, principia rei significat, à quibus sumitur postea operationis initium. Quando autem finem dicit esse principium, cognitionis principium intelligit, quia à notione finis ad principia rei invenienda et cognoscenda progredimur.

8   Quòd si etiam finem diceremus esse principium rei, nil sequeretur absurdi, causae namque sibi invicem causae sunt, satis est, si dicamus Aristotelem, quando ordinem resolutivum inquit esse ad principia, principiorum nomine non intelligere causam finalem, sed materialem, vel effectricem, nam à notione finis ad talem causam inveniendam procedimus ordine resolutivo.

9   Ex his colligimus posse in hoc sensu admitti nonnullorum sententiam, qui dicunt dari ordinem resolutivum à priori, quandoquidem finis, quatenus est causa, est secundùm naturam prior iis, quae ipsius gratia sunt, ordo igitur resolutivus, quatenus à fine exordium sumit, potest dici à priori. In hoc tamen sensu non rectè illi loquuntur, dum dicunt dari aliquem ordinem resolutivum, qui est à priori, cùm potius dicendum sit omnem ordinem resolutivum hoc modo esse à priori. Verùm hi, qui dogma hoc excogitarunt, cùm in alio sensu à priori esse hunc ordinem intellexerint et aliud

It has to be noted, however, lest we suspect that there is an in- 6
consistency in what Aristotle said, that in that passage he says
resolutive order is to beginning-principles, but in the fifth chapter
of the sixth book and in the eighth of the seventh[102] he appears to
assert the contrary, saying that in actions the end is the beginning-
principle (*principium*), just as suppositions [are] in mathematics.
Therefore resolutive order, since it begins from the end, is from
beginning-principles, if the end is the beginning-principle.

But Aristotle does not contradict himself, for when he says that 7
resolutive order is from beginning-principles, he is indicating the
beginning-principles of the thing, from which the start of practical
activity is afterward taken. And when he says that the end is
the beginning-principle, he understands the beginning-principle of
knowledge, because we progress from a notion of the end to discovering and knowing the beginning-principles of the thing.

Now even if we were to say that the end is the beginning- 8
principle of the thing, nothing absurd would follow. For causes
are causes to each other in turn.[103] It is enough if we say that Aristotle, when he says that resolutive order is from beginning-
principles, does not understand by the name "beginning-principles
(*principia*)" the final cause, but the material or efficient [cause]. For
by using resolutive order we proceed from a notion of the end to
discovering such a cause.

From all this we gather that it is in this sense that one can ad- 9
mit the position of some who say that there is resolutive order *à
priori*, since the end, insofar as it is a cause, is according to nature
prior to those things that are for the sake of it. Resolutive order,
therefore, insofar as it takes [its] beginning from the end, can be
said to be *à priori*. Nevertheless, they do not speak correctly in this
sense, when they say that there is some resolutive order that is *à
priori*, since it has to be said rather that every resolutive order is in
this way *à priori*. But those who imagined this doctrine, since they
understood this order to be *à priori* in another sense and regarded

causae genus, quàm finem, respexerint, in magnum et gravissimum
errorem lapsi sunt, quid enim sit ordo resolutivus, se prorsus nes-
cire declararunt. Horum autem et aliorum plurium errores diligentius considerare et expendere operae pretium non est. Satis est ipsam rei veritatem diligentissimè declarasse. Ea enim cognita poterit
per se° quisque parvo negotio aliorum errores deprehendere.

: XVI :

In quo definitio ordinis compositivi
ex dictis colligitur.

1 Utriusque ordinis natura, quae à diversis disciplinarum conditionibus emanat, satis per ea, quae diximus, declarata esse videtur, ut
facile sit ipsorum definitiones colligere, et prius quidem ordinis
compositivi, quem dicimus esse logicum instrumentum, quo cuiusque contemplatricis scientiae partes ita disponimus, ut à primis
rei principiis exordiendo et ad secunda transeundo tandem ad
proxima perveniamus, ut quantum in eo genere fieri possit optimè
et facillimè rerum tractandarum scientiam adipiscamur.

2 Est quidem omnis ordinis communis conditio facere ut facillimè et optimè discamus, sed ad scientiam tanquàm ad ultimum
finem dirigi proprium est ordinis compositivi, qui ab ipsa contem-
plativarum scientiarum natura derivatur.

3 Porrò hic processus à primis principiis ad secunda et ab his ad
tertia nullam significat ratiocinationem, sed solam ordinatam tractationem omnium causarum rei, qui necessariò est processus à

the genus of cause to be other than an end, lapsed into a very large and grave error. For they made clear that they were utterly unaware of what resolutive order is. But to consider and weigh very carefully their and others' many errors is not worth the work. It is enough to have clarified most carefully the very truth of the issue. For, once this is known, anyone on their own° and with little effort will be able to apprehend the errors of others.

: XVI :

In which the definition of compositive order is gathered from what has been said.

It appears that the nature of each of the orders that flows out from   1
the different characteristics of disciplines has been clarified enough by means of what we have said that it is easy to gather the definitions of them—and first, of course, of compositive order, which we say is the logical instrument by which we so dispose the parts of any contemplator's science that by beginning from the first beginning-principles of the thing and going on to the second, we finally move through to the proximate [ones], so that, as much as can be done in the genus, we easily and optimally obtain scientific knowledge of the things being treated.

It is a common characteristic of every order, of course, to enable   2
us to learn easily and optimally. But to be directed to scientific knowledge as to an ultimate end is proper to compositive order, which is derived from the very nature of the contemplative sciences.

This proceeding, moreover, from first beginning-principles to   3
second, and from these to third, indicates no ratiocination but only an ordered treatment of all the causes of something; this is

magis universalibus ad minùs universalia, etenim agere de aliqua re in scientia contemplativa est agere de eius principiis et agere de principiis rei est de illa re agere, cuius sunt principia. Ut in scientia naturali agere de corporibus naturalibus est agere de principiis et de affectionibus corporum naturalium, quemadmodum etiam agere de principiis, vel de affectionibus corporum naturalium est de ipsis corporibus naturalibus agere. Principia verò alia remotiora sunt, alia propinquiora, ut animalium principium remotissimum est materia prima, elementa verò principium propinquius ac veluti secunda materia, demum partes instrumentales sunt materia proxima, de his ergo principiis ordinatè tractare est procedere à magis universalibus ad minùs universalia, id est à genere ad species, nam prima materia est quidem materia hominis et equi, non tamen quatenus est homo et quatenus est equus sed quatenus naturalia corpora sunt, ideò eam materiam considerare est versari in subiecto corpore naturali amplissimè accepto, à quo et homo et equus et alia corpora naturalia tanquàm species à genere continentur. Deinde consideratio elementorum prout sunt secunda materia, est tractatio de corpore misto, cuius principia sunt. Et ita deinceps donec ad principia proxima pervenerimus, quorum consideratio est animalis consideratio quatenus est animal vel hominis quatenus est homo et sic de aliis. Hic igitur idem processus, dum subiectum ipsum respicimus, est à magis universali ad minùs universale et à genere ad species, à corpore naturali ad mistum, à misto ad vivens, à vivente ad animal, ab animali ad hominem et hunc significavit Aristoteles in quarto contextu primi libri Physicorum.

necessarily a proceeding from more universal to less universal. And indeed to deal with some thing in contemplative science is to deal with its beginning-principles, and to deal with the beginning-principles of something is to deal with that thing whose beginning-principles they are. As in natural science, to deal with natural bodies is to deal with beginning-principles and affections of natural bodies, just as also, to deal with beginning-principles or affections of natural bodies is to deal with natural bodies themselves. Now some beginning-principles are more remote, some closer: the most remote of the beginning-principles of animals is first matter; and the next nearer beginning-principle the elements, as if second matter; lastly, the instrumental parts are the proximate matter. Therefore to treat these beginning-principles in order is to proceed from more universal to less universal, that is, from genus to species. For first matter is, of course, the matter of man and horse, not however insofar as it is man or insofar as it is horse, but insofar as they are natural bodies. And so to consider this [first] matter is to be concerned with the subject *natural body* accepted in the widest sense, in which man, horse, and other natural bodies are contained, as species in genus. Then the consideration of the elements, in that they are second matter, is a treatment on mixed body, whose beginning-principles they are. And so on until we have passed through in succession to the beginning-principles, consideration of which is consideration of animal insofar as it is animal, or of man insofar as he is man, and so on with others. When we regard the subject itself, therefore, this same proceeding is from more universal to less universal, and from genus to species, from natural body to mixed, from mixed to living, from living to animal, from animal to man. And Aristotle indicated this in text no. 4 of the first book of the *Physics*.[104]

4   Dum autem principia subiecti spectamus, est à primis ac remotissimis ad minùs remota et secunda, deinde ad tertia donec ad
ultima et propria, quae cuiusque speciei propria sunt, perveniamus.
In qua dispositione[59] consistit ordo compositivus, qui est à simplicibus ad composita, ut à materia prima, quae simplex est, ad materiam secundam, quae constat ex prima; et à prima forma, quae
simplex est, ad secundam, quae est modo quodam composita, ut
ait Averroes in primo commentario primi libri Physicorum. Hanc
ordinatam causarum tractationem secundùm ordinem compositivum significavit Aristoteles in ipso initio primi Physicorum, dum
dixit à primis principiis auspicandum esse, quia ex eorum cognitione omnium aliorum cognitio pendet, hoc enim dicere est ordinem constituere à causis remotis ad proximas, qui est ordo compositivus.

5   Sunt etiam qui putent Aristotelem clarius hunc ordinem in
causarum consideratione ibidem significasse, quando dixit, 'tunc
arbitramur cognoscere unumquodque, quando prima principia et
primas causas noverimus° et usque ad elementa,' quasi principiorum nomine prima principia intellexerit, nomine autem causarum
secunda principia et nomine elementorum principia proxima; sive
per principia et causas intellexerit prima principia, per elementa
verò ultima, proxima. Qua in re Averroes et alii complures decepti
sunt, ea enim non est mens Aristotelis, qui in toto illo contextu
tribus illis vocabulis eandem rem intelligit, nempè prima ac remotissima principia, quae et primae causae et maximè propriè etiam
elementa appellantur.

6   Hoc demonstrare in praesentia intempestivum esset. Aliàs
rem hanc diligentissimè declaravimus, dum locum illum publicè

When, however, we look at beginning-principles of the subject,   4
it is from first and most remote to less remote and second; we then
go through to the third and finally to things ultimate and proper,
what are proper to each species. In this disposition consists compositive order, which is from simples to composites, as from first
matter, which is simple, to second matter which is composed out
of the first, and from first form, which is simple, to second, which
is composite in some way, as Averroës says in the first commentary
to the first book of the *Physics*.[105] Aristotle indicated that this ordered treatment of causes [is] according to compositive order at
the very start of the first [book] of the *Physics*, when he said one
has to commence with first beginning-principles, because knowledge of all others depends on knowledge of these. And to say this
is to establish an order from remote to proximate causes; this is
compositive order.

There are also some who hold that Aristotle clearly indicated   5
this order in the consideration of causes in the same place, when
he said, "We then think that [we] know something when we
come to know° first beginning-principles and first causes, all
the way to the elements,"[106] as if he understood first beginning-
principles by the name of beginning-principles, then second
beginning-principles by the name of causes, and proximate
beginning-principles by the name of elements; or, [as if] by
"beginning-principles and causes," he understood first beginning-
principles, and ultimate and proximate by "elements." On this issue Averroës and many others were deceived, for this is not what
Aristotle had in mind. In that whole text he understands the same
thing in those three terms, that is, the first and most remote
beginning-principles, which are called both first causes and also,
most properly, elements.

Now it would be untimely to demonstrate this at present. We   6
clarified this issue most carefully elsewhere, where we publicly

interpretaremur et ostendimus nullum in eo contextu verbum ab Aristotele proferri, quod proximas causas significet. Sed primas tantùm, ac remotissimas expressit; secundae ac tertiae non exprimuntur, sed tamen subintelliguntur. Dum enim dicit esse primo loco agendum de primis principiis, quae elementa vocantur, significat ab his postea ad secunda, ac tertia esse transeundum.

7   Ad declarandam igitur ordinis compositivi naturam haec, quae hactenus dicta sunt, sufficiant.

: XVII :

In quo definitio colligitur ordinis resolutivi.

1   Ordo autem resolutivus est instrumentum logicum disponens, quo à notione finis, qui ab homine liberè operante produci et generari queat, progredimur ad invenienda et cognoscenda principia, ex quibus operationem postea inchoantes producere et generare finem illum possimus.

2   Est quidem omnium ordinum communis conditio ut ad cognoscendum aliquid conferant,[60] sed ad cognitionem conducere, quae postea in operationem tanquàm ultimum finem dirigatur, proprium est ordinis resolutivi, quandoquidem ex operatricibus disciplinis hic ordo exoritur et ex earum conditione natura huius ordinis derivatur.

3   In hoc quoque ordine ab universalibus ad particularia proceditur, alia tamen ratione, quàm in ordine compositivo. Semper enim quando de aliqua re agendum est et ratione communi et ratione

commented on that passage[107] and showed that in the text no word that signifies proximate causes is advanced by Aristotle. Rather, he expressly stated only the first and most remote [causes]; the second and third are not expressed but instead implied. For when he says that the first beginning-principles, which are called elements, have to be dealt with in the first place, he indicates that there has to be afterward a transition from these to the second and third.

What has been said up until now, therefore, is sufficient to make the nature of compositive order clear. 7

: XVII :

In which the definition of resolutive order is gathered.

Resolutive order, on the other hand, is a logical instrument [for] disposing, by which we progress from a notion of the end, which can be generated and produced by a man practicing freely, to discovering and coming to know beginning-principles. Afterward, starting practical activity with these, we can produce and generate that end. 1

It is, of course, a common characteristic of all orders that they contribute to knowing something, but to contribute to knowledge that is afterward directed to practical activity as an ultimate end is proper to resolutive order, since this order arises from practical disciplines and the nature of this order is derived from what is a characteristic of them. 2

In this order, also, there is a proceeding from universals to particulars, but in a way other than in compositive order. For always, when some thing has to be dealt with in both a way of reasoning that is common and in a way of reasoning that is proper, the 3

propria, ratio communis rationi propriae anteponenda est, universale namque est nobis notius et cognitu facilius, quàm particularia. Ideò quando per ordinem resolutivum fit progressus à fine noto ad invenienda principia ignota, prius universali cognitione principia inveniuntur et communes quaedam ipsorum conditiones in lucem[61] prodeunt, deinde particularem, atque distinctam eorum cognitionem acquirimus, ut in tradenda arte aedificatoria ex praenotione domus prius quaedam communes materiei conditiones inveniuntur, veluti quòd dura esse debet, quae resistat aestui et imbri. Postea particularem eius notitiam° indagantes invenimus talem materiam esse lateres. Ita medicus prius docet curationem febris moliendam esse per remedia frigida et humida, postea ad particularia descendens declarat quaenam sint ea frigida et humida, quibus uti debeamus.

4   Sic Aristoteles in libris de Moribus postquàm docuit felicitatem esse actionem ex virtute, ad tractationem de virtute se contulit et prius quidem[62] egit de virtute generaliter, postea de virtutibus singulis.

5   Ita in Posterioribus Analyticis ex definitione scientiae, ac demonstrationis prius docuit quasdam universales principiorum conditiones in capite secundo primi libri, deinde in capite quarto ad alias magis particulares descendit. Et prius docuit in primo libro principia esse causas, postea in secundo libro distinxit genera causarum, et ostendit quomodo singulum in demonstratione medium accipi possit.

6   Prius itaque à notione finis communem quandam et confusam concipimus principiorum notionem, postea distinctius et particularius ipsorum principiorum notitiam° adipiscimur.

common reasoning has to be placed before the proper reasoning, for the universal is more known to us and easier to know than [are] particulars. And so when progression occurs by means of resolutive order from the known end to discovering the unknown beginning-principles, the beginning-principles are discovered first by universal knowledge, and some of their common characteristics come to light; then we acquire particular and distinct knowledge of them. As in conveying the art of building from the prior notion of house, some common characteristics of the material are discovered first, such as that it ought to be hard, so as to resist heat and rain. Afterward, tracking down particular knowledge° of it, we discover that brick is such a material. So also the physician first teaches that the cure of fever has to be realized by means of cold and moist remedies. Afterward, descending to particulars, he makes clear which are the cold and moist things that we ought to use.

So Aristotle in the *Ethics*, after he taught that felicity is acting out of virtue,[108] devoted himself to a treatment on virtue and dealt first, of course, with virtue generally and afterward with each of the virtues. 4

So also, in the *Posterior Analytics*, from the definition of scientific knowledge and of demonstration, he first taught, in the second chapter of the first book,[109] some universal characteristics of beginning-principles and then, in the fourth chapter, descended to others more particular.[110] And first, in the first book, he taught that beginning-principles are causes;[111] afterward, in the second book, he distinguished genera of causes and showed how each can be accepted as the middle [term] in a demonstration.[112] 5

So first, from a notion of the end, we conceive some sort of common and confused notion of the beginning-principles; afterward we obtain more distinctly and particularly knowledge° of those beginning-principles. 6

: XVIII :

In quo dubium quoddam solvitur de ordine resolutivo.

1   Caeterum non leve dubium oritur ex iis, quae de ordine resolutivo dicta sunt, cuius etiam antea, cùm de ordine in universum loqueremur, meminimus. Ibi namque ordinem à methodo seiungentes diximus in ordine nullam fieri illationem alicuius ex aliquo, sed methodi id proprium esse, hoc tamen falsum esse videtur, quia in ordine resolutivo fit ex necessitate illatio et ratiocinatio, ex fine enim praecognito colliguntur ea, quae ad finem conferunt, ut in primo libro Posteriorum Aristoteles ex definitione demonstrationis colligit per syllogismum qualia debeant esse demonstrationis principia; sic medicus ex definitione sanitatis colligit quibus remediis eius conservatio, vel recuperatio molienda sit.

2   Ad hoc respondere possumus, differentiam inter ordinem et methodum non in eo esse constitutam ut methodus sit semper ratiocinatio, ordo verò nunquàm; sed in hoc quòd methodus necessariò et essentialiter ratiocinatur et illationem facit, haec enim est methodi natura. Ordo autem non semper neque necessariò; non tamen ita, ut ei repugnet, satis est igitur si non omnis ordo sit cum illatione, compositivus enim nullam illationem facit, quòd si resolutivus ex notione finis aliquid colligit, id provenit ex propria natura ordinis resolutivi, non ex natura ordinis, quemadmodum esse risibile competit homini ex propria hominis natura, non ex natura animalis, quae in eo inest. Animali tamen generi risibilitas

: XVIII :

In which some doubt about resolutive order is done away with.

But now another doubt — and not a light one — arises from the 1
things that have been said about resolutive order; we recalled it
earlier when we spoke about order as a universal whole. For there,
separating order from method, we said that no inference of one
thing from another occurs in order, but that is proper to method.
But this appears to be false, because in resolutive order inference
and ratiocination occur out of necessity, for from an end known
beforehand, those things that contribute to the end are gathered.
As Aristotle, in the first book of the *Posterior Analytics,* gathers
from the definition of demonstration [and] by means of a syllogism what sort the beginning-principles of demonstration ought
to be, so the physician gathers from the definition of health the
remedies by which its conservation and restoration are to be realized.

To this we can respond that the difference between order and 2
method is not constituted in that method is always ratiocination
and order never is, but in this, that method necessarily and essentially ratiocinates and makes an inference; this is the nature of
method. Order, however, [does so] neither always nor necessarily,
but it is not such that this [i.e., order] is incompatible with it [i.e.,
ratiocination]. It is enough, therefore, if not every order comes
with an inference, for compositive [order] makes no inference.
And if resolutive [order] gathers something from a notion of the
end, this arises from the nature proper to resolutive order, not
from the nature of order, just as to be risible appertains to man
from the nature proper to man, not from the nature of animal that
belongs to him. Risibility is, nevertheless, not incompatible with

non repugnat, proinde neque omne animal est risibile, neque etiam nullum.

3   Aliam tamen possumus afferre responsionem, quam ego magis probo, multa sunt, quae pro diversis considerationibus diversas naturas prae se ferunt, ut homo, quatenus est homo, sub genere substantiae collocatur; idem quatenus est albus est in genere qualitatis; at quatenus est pater est in numero eorum, quae dicuntur ad aliquid. Quod quidem evenit propterea quòd substantia alias categorias coniunctas habet.

4   Sic dicimus de ordine resolutivo, methodum enim necessariò habet coniunctam et alia ratione dicitur ordo, alia ratione methodus. Facit quidem illationem, quatenus est methodus, cuius est proprium ratiocinari, at quatenus est ordo nil aliud facit, quàm dispositionem; quòd enim prius de fine agatur, postea de iis, quae ad finem ducunt, id ei competit quatenus est ordo resolutivus, quòd verò ex fine colligantur illa,[63] quae ad finem conferunt, competit ei ratione methodi, quae cum eo est coniuncta, quae qualisnam methodus appellanda sit, postea in tractatione de methodis considerabimus.

5   Hanc autem differentiam animadvertere possumus in libro Averrois, qui dicitur Colliget, ab eo doctissimè declaratam, dum separatim et ordine resolutivo et methodo illa utitur, quae cum eo connexa esse solet. Nam à notione finis ad remedia ascendens voluit prius in quinto libro de ipsis remediis secundùm se tractationem facere et de ciborum ac medicamentorum natura disserere sine ulla illatione ex fine. Dum igitur librum illum cum praecedentibus comparamus, dicimus[64] Averroem prius de fine locutum esse, postea de principiis absque ulla illatione horum ex illo, quod quidem est ordine uti absque aliqua methodo. In duobus autem sequentibus ac postremis libris usum methodi adhibet. Quemlibet

the genus of animal; and, accordingly, neither is every animal risible, nor is none.

We can, however, bring forward another response, one which I   3
much approve of. There are many things that, on account of different considerations, exhibit different natures, as [for example] man insofar as he is man is included under the genus of substance, and the same insofar as he is white is in the genus of quality, and insofar as he is a father is in the number of those that are said to be [relative] to something. This happens, of course, because of the fact that substance has other conjoined categories.

We speak thus of resolutive order. Now method has a conjunc-   4
tion necessarily,[113] and in one way of reasoning is called order and in another way of reasoning method. It makes an inference, of course, insofar as it is method, to which ratiocinating is proper, but insofar as it is order, it makes nothing other than a disposition, in that it deals first with the end and afterward with those things that lead to the end. This appertains to it insofar as it is resolutive order, but that those things that contribute to the end are gathered from the end appertains to it by reason of method, which is conjoined with it. What a method of this sort is to be called, we will consider below in a treatment on methods.

We can note this difference in Averroës' book that is called *Col-*   5
liget, where it is made clear by him in a most learned way, when he separately uses both resolutive order and the method that is normally connected to it. For, ascending to remedies from a notion of the end, he wanted first, in the fifth book, to make a treatment on remedies just in and of themselves and to discuss the nature of nutriments and medicaments without any inference from an end. When we compare this book with the preceding ones, therefore, we say that Averroës spoke first about the end, afterward about the beginning-principles without any inference of them from that [end]; this, of course, is to use order without any method. In the two following and final books, however, he applies the use of

enim morbum expellendum vel sanitatem conservandam proponens colligit, quaenam remedia adhibenda sint. Prius igitur Averroes ordinem servat, dum prius de fine agit, postea de medicamentis; deinde verò methodum, quando haec ex illo colligit, quare
nullum est logicum artificium, quod in libro illo Averrois non
inspiciatur.

6   Quae igitur Averroes reipsa distinxit, nos in aliorum libris ordine resolutivo scriptis, si reipsa distincta non sint, ratione distin-
guamus et dicamus eundem processum alia ratione ordinem esse,
alia ratione methodum, sic enim omne dubium, ni fallor, sublatum
erit.

∶ XIX ∶

De ordinum inter se comparatione.

1 Quaerere aliquis posset uter duorum ordinum utilior et uter nobilior seu praestantior sit, nam et alii, qui de his tractationem insti-
tuerunt, huiusmodi ordinum collationem° fecerunt.

2   Attamen nos absurdam esse hanc comparationem et hanc
quaestionem arbitramur, et rationi minimè consentaneam. Nam
locum quidem ea quaestio haberet apud eos, qui dicunt unam et
eandem disciplinam ordines omnes admittere, sic enim quaeri
posset,[65] an in tradenda scientia naturali utilior hic ordo sit, an ille
et uter praestantior sit. Sed cùm ordo compositivus solis scientiis
contemplativis conveniat, resolutivus autem solis operatricibus

method. For, in setting out any disease to be expelled or health to be conserved, he gathers what remedies are to be applied. Averroës, therefore, first maintains order, when he deals first with the end and afterward with the medicaments, but then [maintains] method when he gathers the latter from the former. There is, therefore, no logical skill that is not observed in that book of Averroës'.

What Averroës distinguished in reality, therefore, let us, in the 6 books of others written using resolutive order, distinguish in reasoning (if they are in reality not distinct) and let us say that the same procedure is in one way of reasoning order and in another way of reasoning method. And in this way every doubt, unless I am mistaken, will be removed.

: XIX :

On comparison of the orders to each other.

Someone could inquire which of the two orders is more useful and 1 which is more noble or more excellent. For others, too, who put together a treatment on this made a comparison° of orders of this type.

But we, nevertheless, think that this comparison and this ques- 2 tion are absurd and do not agree with reason at all. Now this question, of course, would have had a place in those who say that one and the same discipline admits all orders. For then it could be inquired whether, for conveying natural science, this order or that one is more useful, and which of the two is more excellent. But since compositive order is appropriate only to contemplative sciences, and resolutive only to practical disciplines, this comparison

disciplinis, certè comparatio haec absurditate non caret, quemadmodum si quis quaereret utrum sit nobilius vel utilius instrumentum serra fabri an malleus aedificatoris; sive uter eruditior fuerit,
Averroes an Bartholus. Dum enim scientias contemplativas consideramus, dicere non possumus ordinem compositivum resolutivo
utiliorem esse, sed dicendum est compositivum utilem esse, resolutivum autem prorsus inutilem. Contra verò in aliis disciplinis
resolutivum solum esse utilem, compositivum inutilem. Quare
proportione quadam aequè utiles dici possunt, ut enim ad contem-
plativarum scientiarum traditionem compositivus, ita ad aliarum
disciplinarum resolutivus utilis est.

3   Ordinum autem nobilitatem aliunde, quàm ex ipsarum disciplinarum nobilitate desumere ridiculum est; nam secundùm se nullam maiorem praestantiam hic ordo habet, quàm ille, sed quatenus
scientiae contemplativae aliis omnibus disciplinis nobilitate prae-
stant, eatenus ordo compositivus resolutivo praestantior est.

4   Videtur tamen id quoque posse considerari, quòd ordo compositivus nunquàm ad resolutivum dirigitur. Resolutivus autem semper est propter compositivum, tanquàm propter finem, proinde
nobilior est compositivus, finis enim iis, quae sunt ante finem,
praestantior est. Quòd autem resolutivus ad compositivum dirigatur, manifestum est, ducit enim ad principiorum cognitionem, à
quibus postea operatio incipiat, quae procedit ordine compositivo.
Sed revera hic non est ordo ille compositivus, de quo in praesentia
loquimur. Ordo enim, de quo sermonem habemus, est instrumentum intellectuale ad cognitionem conferens, quare ordo compositivus sive artis sive naturae operantis, ad nostram considerationem
non pertinet, nisi quatenus ipsius mentionem facere coacti sumus
ad declarandam resolutivi ordinis naturam.

5   Dum igitur et compositivum et resolutivum ordinem accipi-
mus prout instrumenta logica sunt ad disciplinarum traditionem,

certainly does not lack absurdity—just as if someone were to inquire which is a more noble or more useful instrument, the work-
man's saw or the builder's hammer, or who was more learned,
Averroës or Bartolus.[114] For when we consider contemplative sciences, we cannot say that compositive order is more useful than
resolutive; rather, it has to be said that compositive is useful and
resolutive utterly useless. On the other hand, in other disciplines,
resolutive [order] alone is useful, compositive useless. Therefore
they can be said to be equally useful in some sort of comparative
relation. For as the compositive is useful for the conveying of contemplative sciences, so resolutive [is] for other disciplines.

So to draw the nobility of orders from other than the nobility   3
of the disciplines themselves is ridiculous; for in and of itself this
order has no greater excellence than that. But insofar as contemplative sciences excel in nobility all other disciplines, so compositive order is more excellent than resolutive.

Moreover, it also appears that it can be considered that com-   4
positive order is never directed toward the resolutive. Resolutive,
however, is always for the sake of the compositive, as for an end.
And so compositive is more noble, for the end is more excellent
than the things that are before the end. Also, that resolutive [or-
der] is directed toward compositive is manifest, for it leads to
knowledge of beginning-principles, from which practical activity
afterward starts, [activity] that proceeds using compositive order.
But in truth this is not the compositive order about which we are
speaking at present. For the order that we are discoursing on is
an instrument of [the] understanding contributing to knowledge.
And so the compositive order either of an art or of the nature of
practicing does not pertain to our consideration, except insofar as
we are forced to make mention of it to make the nature of resolutive order clear.

While we accept, therefore, both compositive and resolutive   5
order in that they are logical instruments for the conveying of

proinde ad cognitionem conferentia, nec ordo compositivus est propter resolutivum, nec resolutivus propter illum compositivum, de quo in praesentia sermo est. Remanet igitur illa una ordinum comparatio, quae in ipsarum disciplinarum inter se comparatione consistit.

: XX :

De ordine universali et particulari.

1 Non est autem silentio praetereundum aliud esse ordinem universalem, aliud esse ordinem particularem, universalem illum vocamus, quo totam disciplinam disponimus; particularem verò eum, qui in aliqua disciplinae parte servatur. Nos quidem de universali tantùm ordine locuti sumus, quia de ordine particulari nulla certa praecepta tradi possunt praeter id, quod de ordine universè diximus, nempè illa esse anteponenda, quorum cognitio ad reliquorum cognitionem necessaria sit. Evenire autem potest in parte alicuius scientiae traditae ordine compositivo ut ordo aliquis servetur, qui per se non sit compositivus; et in parte alicuius artis traditae ordine resolutivo aliquis ordo servetur, qui non sit resolutivus; quam quidem rem exemplis facilè declarabimus.

2 Si totam scientiam naturalem spectemus, eam dicimus scriptam esse ordine compositivo, quoniam à primis principiis auspicatur et in ultimis ac proximis desinit. Si verò partem aliquam eius scientiae consideremus, ut octo libros Physicae auscultationis, eamque ad totum referamus, adhuc dicimus eam ordine compositivo scriptam esse, cùm enim hic sit totius scientiae ordo, est etiam

disciplines, and thus as contributing to knowledge, neither is compositive order for the sake of resolutive nor resolutive for the sake of that compositive which the discourse at present is about. There remains, therefore, that one comparison of orders that consists in a comparison of the very disciplines to each other.

: XX :

On universal and particular order.

It should not, however, be passed over in silence that universal   1
order is one thing, and particular order is another. We call universal that by which we dispose the whole discipline, and particular that which is maintained in any part of the discipline. Of course, we have spoken only about universal order, because no definite precepts can be conveyed about particular order besides what we said about order universally, namely that those things have to be placed first, knowledge of which is necessary for knowledge of the rest. It can happen, however, in a part of some science conveyed using compositive order, that some order is maintained that is not *per se* compositive, and in a part of some art conveyed using resolutive order, some order is maintained that is not resolutive; we will clarify this easily, of course, using examples.

If we look at natural science as a whole we say it is written us-   2
ing compositive order, since it commences with first beginning-principles and finishes in ultimate and proximate [ones]. But if we consider some part of this science, as [for example] the eight books of the *Physics Lectures* [i.e., the *Physics*], and refer it to the whole, we still say it is written using compositive order. For although this is the order of the science as a whole, it is also the

ordo omnium partium prout totius partes sunt et totum ipsum constituunt.

3   At si ipsam secundùm se partem sumamus sine relatione ad totum, alia ratio est. Possumus quidem in libris illis dicere ordinem servari compositivum quatenus de principiis corporis naturalis prius agitur, postea de ipsius accidentibus.

4   Sed si primum solum librum accipiamus, in quo de primis principiis agitur, ille ordinem quidem habet compositivum, quatenus est prima pars et initium ordinis compositivi, at per se sumptus neque compositivum neque resolutivum ordinem habet, quia in eo non plures tractationes, sed una tractatio fit, quae est de primis principiis. Plura quidem theoremata declarantur attinentia ad ipsa prima principia et in illis hic ordo servatur, quòd prius theoremata magis confusa et universalia, deinde magis distincta et particularia explicantur. Sed hic non magis est ordo compositivus, quàm resolutivus, cùm in utroque à confusis ad distincta et ab universalibus tum rebus tum quaestionibus ad eandem rem attinentibus ad particulares progressio fiat, qui potest vocari ordo compositivus, quando est pars ordinis compositivi, et resolutivus, quando est pars resolutivi.

5   Eadem ratione si quis quaerat an liber octavus Physicorum sit scriptus ordine compositivo an resolutivo, inanis quaestio est. Nam compositivus vel[66] resolutivus ordo servari dicitur, quando multae res tractantur una post aliam. Sed in eo libro non de pluribus rebus, sed de una re agitur, de aeterno motu, ad quem unum demonstrandum plura theoremata discutiuntur ab Aristotele, quae ita disponuntur, ut confusa distinctis anteponantur, quemadmodum etiam de primo libro diximus. Ibi itaque qualisnam sit methodus Aristotelis, annotare possumus. Methodus enim est instrumentum

order of all the parts in that they are parts of the whole and constitute the whole itself.

But if we take a part just in and of itself, not in relation to the 3 whole, the reasoning is otherwise. We can say, of course, that in these books compositive order is maintained, insofar as the beginning-principles of natural body are dealt with first, and afterward its accidents.

But if we accept just the first book, in which the first beginning- 4 principles are dealt with, it has, of course, compositive order, insofar as it is the first part and the start of compositive order. But taken *per se*, it has neither compositive nor resolutive order, because in it only one treatment is made, not many treatments, [and] it is on first beginning-principles. Many theorems, of course, pertinent to first beginning-principles themselves are made clear, and in them this order is maintained: theorems more confused and universal are explicated first, and then those more distinct and particular. But this is no more compositive order than resolutive, since in each of the two, progression occurs to distinct from confused and to particulars from universals, both [universal] things and [universal] questions pertinent to the same thing. This can be called compositive order when it is a part of compositive order, and resolutive when it is a part of resolutive.

For the same reason, if someone were to inquire whether the 5 eighth book of the *Physics* is written using compositive or resolutive order, the question is inane. For compositive or resolutive order is said to be maintained when many things are treated, one after another. But in this book not many things but [just] one thing, eternal motion, is dealt with. To demonstrate this one thing, many theorems are struck down by Aristotle. These are so disposed that the confused are placed before the distinct, just as we said too about the first book. And so we can note what sort of method Aristotle uses there. For method is the instrument of the

partis, sed ordo est instrumentum totius, quia est ordinata multaram rerum tractatio in tota disciplina. Unius autem rei tractatio ordinem eatenus habere dicitur compositivum vel resolutivum, quatenus ad alias refertur tum prius consideratas tum posterius considerandas.

6   Aliqua etiam in scientia naturali pars est, in qua ordinatim multa tractantur, tamen nec ordo compositivus est, nec resolutivus, ut inter libros Physicae auscultationis illi, in quibus agitur de communibus accidentibus corporis naturalis, scilicet de motu, de loco ac de tempore, prius enim de motu sermo fit, postea de loco ac de tempore, deinde iterum de motu; locus enim est natura prior motu, tempus autem est posterius, quare cùm motus loco et tempori tum anteponatur tum postponatur, nullus certus ordo servatur, nisi facilioris nostrae cognitionis. Hoc autem adhuc evidentius est ubi plura eiusdem subiecti accidentia tractantur, quorum nullum est natura prius alio neque unum ex alio pendet, eorum enim dispositio non potest vocari ordo compositivus neque resolutivus secundùm se, tamen respectu totius scientiae est ordo compositivus, quia est pars quaedam ordinis compositivi totius naturalis philosophiae.

7   In artibus quoque idem possumus annotare, ars enim medica tradita est ordine resolutivo, sed prima eius pars, quae physiologica dicitur, si ipsa secundùm se consideretur, non solùm non habet ordinem resolutivum, sed potius contrarium compositivum, quod multos in eum errorem traxit, ut crediderint artem medicam traditam esse ab Avicenna ordine compositivo. Sed non animadverterunt aliud esse artem medicam considerare, aliud esse primam illam partem secundùm seipsam sumere absque ullo respectu ad alias medicinae partes. Ars enim medica traditur ordine resolutivo et pars eius physiologica ordinem habet resolutivum, quatenus

part, but order is the instrument of the whole, because it is a treatment of many things, ordered in the whole discipline. A treatment of one thing, however, is said to have compositive or resolutive order only insofar as it refers to others, both to what has been considered before and to what is to be considered after.

In natural science there is also some part in which many things 6 are treated in an order, but nevertheless the order is neither compositive nor resolutive, as [for example] those [parts] in the books of the *Physics Lectures* [i.e., the *Physics*], in which the common accidents of natural body, namely motion, place, and time, are dealt with. For first, there is discourse about motion, afterward about place and about time, then about motion again—for place is by nature prior to motion and time is posterior. And so, since motion is placed both before and after place and time, no definite order is maintained, except [that chosen for] our easier knowledge. Now this is even more evident, where many accidents of the same subject are treated, of which none is by nature prior to another, nor does one depend on another. The disposition of these cannot be called compositive or resolutive order in and of itself, but with regard to the whole science it is compositive order, because it is some part of the compositive order of natural philosophy as a whole.

We can note the same thing also in the arts. For medical art is 7 conveyed using resolutive order, but its first part, which is said to be physiology, if it is considered just in and of itself, not only does not have resolutive order, it has instead the contrary, compositive. This drove many into the following error: they believed that the medical art was conveyed by Avicenna using compositive order. But they did not notice that it is one thing to consider the medical art, and it is another to take that first part just in and of itself, without any regard to the other parts of medicine. For medical art is conveyed using resolutive order, and its part, physiology, has resolutive order, insofar as it is a part of the whole art, and regards

totius artis pars est et respicit alias partes ipsamque artem totam, est enim pars prima et initium ordinis resolutivi tradens nobis notionem finis, quemadmodum antea declaravimus. At si secundùm se pars illa consideretur, non prout est illius artis pars et alias partes respicit, ordinem quidem habet compositivum, sed non est amplius ars medica, neque ipsius pars, sed scientiae naturalis, quia contemplari humani corporis fabricam° sine ullo respectu ad curationem vel aliam operationem, ad quam illa contemplatio dirigatur, officium est philosophi naturalis.

8   Videmus enim in arte medica processum fieri ab universalibus ad particularia, à confusis ad distincta, ut in quinto Averrois libro ostendimus et in secundo Avicennae, qui secundùm se non est ordo resolutivus, neque compositivus, cùm in utroque ordine locum habeat. Tamen respectu totius artis est ordo resolutivus, id est pars ordinis resolutivi.

9   Haec volui annotare propter nonnullos, qui nodum in scirpo quaerunt et putant necessarium esse tum in tota disciplina, tum etiam in qualibet eius parte ordinem servari talem, qui sit aliqua ordinis species, nimirum vel compositivus, vel definitivus, vel resolutivus; quasi necessarium sit, si totus homo est animal, oculum quoque hominis animal esse; atqui oculus est quidem animal, quatenus est pars hominis, qui verè est animal, at ipse per se non est animal: ita omnis artifex totam aliquam disciplinam traditurus certo quodam ordine res in ea tractandas disponit, qui vel compositivus ordo est vel resolutivus, in singula autem parte non semper ordinem servat compositivum vel resolutivum, sed ordinem tamen servat, dum ea prius declarat, quae ad aliorum declarationem prius cognovisse oportuit, quae est omnis ordinis conditio et ille dicitur ordo compositivus vel resolutivus habita ratione totius, cuius pars est. Participat enim conditione ordinis ut ordinis pars, non ut ordo integer.

the other parts and the whole art itself. For it is the first part and the start of resolutive order, conveying to us a notion of the end, as we made clear earlier. But if that part is considered in and of itself, not as it is a part of that art and [not as it] regards other parts, then it has compositive order, of course, but then it is no longer medical art or part of it but is [a part] of natural science, because to contemplate the construction° of the human body, without any regard to curing or other practical activity to which that contemplation is directed, is the role of the natural philosopher.

For we see that in medical art proceeding occurs from universals to particulars, from confused to distinct, as we showed in Averroës' fifth book and in Avicenna's second; in and of itself, this is not resolutive or compositive order, since it has a place in either order. Nevertheless with regard to the whole art, it is resolutive order, that is, a part of the resolutive order. 8

I wanted to note this for the sake of those who seek a knot in a bulrush[115] and hold it necessary both in the whole discipline and also in any part of it that such an order be maintained, such that it is some species of order, namely, of course, either compositive, definitive, or resolutive; as if it were necessary that if the whole man is an animal, then the eye of man is an animal also; but yet the eye, of course, is an animal insofar as it is a part of man, who truly is an animal, but itself *per se* is not an animal. And so every practitioner, about to convey some discipline as a whole, disposes the things being treated in it using some definite order, which is either compositive order or resolutive. He does not, however, always maintain, in every part, compositive or resolutive order. But he nevertheless maintains an order when he first makes clear those things that must become known first, for clarification of other things. This is a characteristic of every order, and this order is said to be compositive or resolutive by taking account of the whole of which it is a part. For it participates in a characteristic of the order, as a part of the order, not as the [whole] order intact. 9

10   Sic etiam oculus sensu participat, in quo tota animalis natura consistit, nec tamen sequitur oculum per se animal esse, non est enim animal neque species animalis ulla, quia conditione animalis participat ut animalis pars, non ut animal totum.

11   Quemadmodum igitur datur aliquod sensu praeditum, quod non est animalis species aliqua, ita datur in disciplinis aliqua dispositio, quae nec est ordo compositivus nec resolutivus, est tamen pars aut huius aut illius.

12   Haec de ordinibus dicta sint, nunc ad methodos transeamus.

And so the eye participates in sense, in which the whole nature   10
of the animal consists, but it does not follow that the eye is *per se*
an animal. For it is not an animal or any species of animal, because
it participates in what is characteristic of animal as a part of an
animal, not as the whole animal.

Just as there is something endowed with sense, therefore, that is   11
not some species of animal, so there is in disciplines some disposition that is neither compositive nor resolutive order, but nevertheless is a part either of the latter or of the former.

Having said this about order, let us now move on to methods.   12

Note on the Text and Translation

ॐ᠖ॐ

The Latin text is based on a full collation of the two editions published in Venice during Zabarella's lifetime,

A   [Editio Prima]. Venice: Paulo Meietti, 1578.
B   Editio Secunda. Venice: Paulo Meietti, 1586.

and two posthumous editions published in Germany,

C   Editio Tertia. Cologne: Lazarus Zetzner, 1597.
E   Editio Postrema. Frankfurt: Lazarus Zetzner, 1608.

A third posthumous German edition was consulted wherever a difference in the other four was discovered:

d   Editio Quarta. Cologne: Lazarus Zetzner, 1603.

These posthumous editions are not linear descendants of either of the Venice editions. They descend presumably from the edition published by Jean Mareschal in 1586–87, which has not been collated.*

The text printed here generally follows B. The differences, though small, between it and A suggest the changes were made by

---

* Maclean, "Mediations of Zabarella," 47, states: "It can be deduced also from the same preface that Pace did not consult the author himself in advance about the publication of his works" [including the *Opera logica*]. The list of editions in the Bibliography is heavily indebted to Maclean's study. Copies of the Mareschal 1587 edition can be found at the Huntington Library and at the Bayerische Staatsbibliothek; the latter has made the edition available online in its digital collection.

the author himself. Punctuation here also follows the second edition, although some discretion has been exercised. None of the consulted editions had paragraph breaks; these have been added in keeping with the editorial practice of the I Tatti Renaissance Library. Generally, these breaks follow what look typographically in the source like boundaries between very long sentences. Semicolons in the Latin have been changed to periods where the English translation warrants it. Serial commas were regularly used in lists of only two items, and these have been silently removed.

In the sources, *et* was always rendered as an ampersand; it has here been expanded. The periods after Arabic numerals and after capital letters when used as symbols have been removed. The way numbers were presented in the source texts varied and these have been silently rationalized.

In the sources, the conjunction *quum* was always distinguished from the preposition *cum*. The first has been rendered here as *cùm*. Grave accents, such as those distinguishing *quòd* (adverb, conjunction) from *quod* (pronoun), *quàm* (adverb, conjunction) from *quam* (pronoun), and *secundùm* (preposition) from *secundum* (adjective) have been retained and silently rationalized in the few cases where it was deemed necessary. These distinguishing marks are often helpful in determining the sense and are occasionally disambiguating. An exception has been made for *-iùs/-ius*. Later editions attempted, however inconsistently, to use the accent to distinguish adjectives and adverbs of the comparative. The first edition did not, and that edition's practice has been adopted here.

Some editions used a circumflex to mark the genitive plural in the phrase *nostrûm cognoscentium*. These endings have silently been expanded to *-orum* and the phrase translated as "our knowing." The circumflex was sometimes used to distinguish the adverb *hîc* (here) from the demonstrative *hic* (this) and has here been retained.

Acute accents, used with enclitics, as -*ne*, -*am*, and -*que*, have been removed.

Spellings silently changed include *secu-* for *sequu-*, -*sid-* for -*syd-*, and -*m-* for -*n-* in compounds such as *quanvis* and *nanque*. Alternate spelling, such as *qua propter/quapropter* and *eapropter/propterea* have been silently rationalized. Banal typographical errors have been silently corrected.

Other variations are noted in the apparatus.

THE TRANSLATION

Zabarella's works were part of a tradition of carefully worded philosophical commentary and were part of an international dialogue conducted in multiple languages. To preserve the precision of the commentary and properly situate the texts in that dialogue, the translation here is rather literal, even wooden.

Attempts have been made to retain ambiguities. When insertion of an article could alter meaning and might not actually be warranted, the insertion is marked (e.g., "habit of [the] understanding"). Pronouns whose resolution is apparent in the Latin because of, say, gender, are similarly marked as insertions. Others are left unresolved.

Great care has been taken with vocabulary, with two goals in mind. The first goal is to employ terminology that would not be far from what Zabarella's English readers and discussants used in their own writings on the same subjects. Consequently, the reader of this translation may find old-fashioned uses of some words. For example, "to dispose" here means to arrange, not to discard; a disposition is an arrangement; "vain" means empty. "Speculative sciences" are theoretical ones, as opposed to practical; there is no suggestion that the science is unfounded or merely hypothetical, as the term might nowadays suggest. In the early seventeenth cen-

tury, the translation of *prima materia* was "first matter," it was not yet "prime matter," and so the former is here used. Other words of later origin, especially technical ones such as "deduction," have been avoided.

The second goal is a highly consistent correspondence between Latin and English terms. For example, *cognoscere* is translated "to know" (as in Zabarella's time), even when "to understand" or "recognize" might seem better to our ear. "To understand" is reserved for *intelligere*. *Ratio* has a few regular translations, chosen based on context: "reason" or "reasoning" when standing alone, "rational way of . . ." in phrases such as "rational way of ordering." Cognates are common, such as "manifest" for *manifestus*, "apparent" for *apparens*. Throughout, exceptions and irregular choices are marked with a superscripted ring on both Latin and English; e.g., *rationibus ita apparentibus° nixi* (they relied on reasons so specious°).

It is hoped that reading the translation here will feel no more unfamiliar than reading a comparable text written by an English logician of the early seventeenth century.

Terms of particular importance include the following.

via	way
methodos	method
inferre	infer
tradere	convey (more literally: deliver, pass on)
ducere/duci	lead/be led
inducere	induce
deducere	deduce

Some associations here that would be plain to Zabarella are inevitably lost in English. Greek *hodos* is a path or way; the compound *meth-odos* a pursuit along a path. Zabarella draws attention to the connection between Greek *methodos* and Latin *via* at *Meth. l. 1 c. 2. par. 1, l. 1 c. 7 par. 4, l. 3 c. 2 par. 2.* For Zabarella, "to infer"

(*inferre*, to carry from point to point) and "to be led" (*duci*) have strong literal associations with "way" and "method," while "deduce" (used very seldom), "induce," and "lead" (used often in the passive) have stronger affinities than the English would suggest. The phrase "method by which we are led to knowledge" (*Meth. l.* 4 *c.* 10 *par.* 3) has for Zabarella a naturalness, even an inescapability, that is easy for us to overlook. A similar association is echoed in his frequent statement that knowledge is conveyed by method.

deducere	deduce
inferre/illatio	infer/inference
concludere/conclusio	conclude, is conclusive, be a conclusion/ conclusion
colligere/collectio	gather/gathering

Of these four, *colligere* is by far the most common and has the widest range of meanings. It is always translated here as "gather" but occasionally — and especially in *Meth. l.* 3 — seems to refer specifically to deductive reasoning. For Zabarella, the conclusion of a syllogism is "gathered" from its premises. And it may be that a *collectio* (gathering) sometimes means what we would now call a "deduction," a much later term. Note also use of the term in *Meth. l.* 4 *c.* 13.

declarare	make clear, clarify
declaratio	clarification

The senses are very close to "explain" and "explanation," but the alternatives used here keep close the association with *clarus* (clear), a property of some mental content and a property central to Zabarella's epistemology (and later Descartes').

ratio	reason, reasoning, rational way (as in *ratio ordinandi*, rational way of ordering)

Zabarella frequently uses *ratio* with a gerund in the genitive, as *ratio ordinandi*. The bare meaning is "the way" or "the manner" of doing something, but "right, proper, and rational" is generally implied. The usage places Zabarella squarely within humanistic discussions of *ratio dicendi, ratio docendi*, etc., the proper manner of speaking, the right way of teaching, etc. Even *ratio* by itself has a sense of "the correct way of reasoning."

disponere	dispose
dispositio	disposition

The sense is "to arrange," not "to discard." "Disposition" is synonymous with "order."

tractare	treat
tractatus	treatise
tractatio	treatment, tract

A *tractatio* is both an examination of some topic and the body of text containing that examination. So it can be a passage, a section of a book, or even a complete book on a specialized topic.

cognoscere	know
noscere°	know°

Holyoake explains that *noscere* means "to know" and *cognoscere* means "to come to know." Zabarella uses the first very seldom and the second very often. Though *cognoscere* is here translated simply as "to know" (exceptions are noted), it should be remembered that the term does carry a sense of "coming to know," rather than just knowing, and indeed the meaning always hovers close to our "recognize." Remembering this is vital for understanding Zabarella's epistemology generally.

cognitio	knowledge
notitia°	knowledge°

In Zabarella's day, "knowledge," not "cognition," was the standard translation for *cognitio*, and the practice is maintained here. It seems to accord well with his meaning (though *cognitio* always carries some sense of "recognition" as well; see above, s.v. "cognoscere"). Zabarella uses *notitia* much less frequently. English "knowledge" is used for both, though instances of *notitia* are marked as exceptions. It is left to the reader to determine whether Zabarella intends a difference in meaning. *Regr. l.* 9 *c.* 5–6 suggest he does not; *Meth. l.* 3 *c.* 1–2 may suggest he does. Compare also his commentary on the *Posterior Analytics l.* 1 *c.* 18 *t.* 134. Schicker used *erkenntnis* for both.

scire	know scientifically, have scientific knowledge
scientia	a science, the science, scientific knowledge
sciendum° est	it has to be understood° that
intelligere/intellectus/ intellectualis	understand/understanding/ of understanding

Zabarella almost always uses *scire* in the narrow sense corresponding to Aristotle's *epistasthai* and *scientia* in the sense of *epistēmē* (or the *epistēmē* of a particular subject). See Zabarella's commentary on *Posterior Analytics l.* 1 *c.* 2 71b9–15, *Comm. Post. An. l.* 1 *t.* 7. "Understand" is reserved for *intelligere* (with exceptions noted), though occasionally the sense is "to understand the meaning to be, to mean."

prior/notior nobis	prior/more known to us
prior/notior natura	prior/more known by nature

prior/notior secundùm naturam	prior/more known according to nature

In the dichotomy *notius natura* vs. *notius nobis*, which translates Aristotle's γνωριμώτερον τῇ φύσει (*gnōrimōteron tēi phusei*) vs. γνωριμώτερον πρὸς ἡμᾶς (*gnōrimōteron pros hemas*), it can appear that both *natura* and *nobis* are ablative. But at least for Zabarella, only the first is; the second is dative. In English the terms are "more known *by* nature" and "more known *to* us."

Moreover, Zabarella criticizes those who took "more known by nature" to indicate that nature was the knower. "Nature," he says, "knows nothing." The phrase should instead be understood to mean known by us in a way "that maintains the order of nature." Seeking, presumably, to distance himself from the common misconception, he generally uses *notior secundùm naturam* (more known according to nature) when speaking in his own voice and *notior natura* (more known by nature) when citing or paraphrasing his predecessors, even though he indicates that the two are synonyms.

For explications of Zabarella's position, see especially his commentary on *Posterior Analytics l. 1 c. 2 71b33–72a7, Comm. Post. An. l. 1 t. 12*, but also *Meth. l. 4 c. 10* and *De Speciebus Demonstrationis c. 17*.

demonstratio quòd	demonstration *quòd*	*quòd*: that something
cognitio quòd	knowledge *quòd*	is the case
notitia° quòd	knowledge° *quòd*	

demonstratio quid est	demonstration *quid est*	*quid est*: what something is
cognitio quid est	knowledge *quid est*	
notitia° quid est	knowledge° *quid est*	

demonstratio propter quid (est)	demonstration *propter quid*	*propter quid*: what something is on account of
cognitio propter quid (est)	knowledge *propter quid*	
notitia° propter quid (est)	knowledge° *propter quid*	
demonstratio à signo	demonstration *a signo*	*a signo*: from an indicator or sign
demonstratio ab effectu	demonstration *ab effectu:*	*ab effectu*: from an effect
demonstratio quia	demonstration *quia:*	*quia*: because
demonstratio potissima	demonstration *potissima*	*potissima*: of the strongest sort
demonstratio simpliciter	demonstation *simpliciter*	*simpliciter*: absolute

As early as Zabarella's own time, *potissima demonstratio* has been translated "perfect demonstration" or "most perfect demonstration," but this is misleading. As Zabarella explains in *Meth. l.* 3 *c.* 4 and *De speciebus demonstrationis* 12, *potissima demonstratio* translates the Greek *kurios apodeixin* and means the same as demonstration *propter quid*. Greek *kurios* means "powerful or strong."

A demonstration *propter quid*, an "on account of which" demonstration, is not a demonstration *of* a cause. The cause is in the premises of the syllogism as a middle term; it is not in the conclusion.

In *Meth. l.* 3 *c.* 4, Zabarella says that demonstration *a signo* and demonstration *quia* are the same and in *l.* 3 *c.* 16 uses *a signo* and *ab effectu* interchangeably. In *Regr.*, he explains that a demonstration *quod* is a demonstration *ab effectu*.

Zabarella uses *demonstratio simpliciter* only in *Regr. c.* 3, where he says it is demonstration from what is better known in accord with nature.

Note, especially for understanding *Meth. l.* 4, that "demonstration" and "demonstrative method" are not synonyms. Only some demonstrations qualify as demonstrative method.

certè	certainly
certus	certain, definite

Care should be taken not to read into the term too much difference between instances translated as "certain" and those translated as "definite." Literally, the word refers to what is seen and discerned. Holyoake says it has extended meanings such as "true," "necessary," and "manifest," "because what we discern is clear and indubitable." Zabarella has no reservation about using the term in the comparative and the superlative. He uses it both for "certain knowledge" and "a definite order," but the meanings are not as different for him as they are for us.

perfectus	perfect

The sense is nearly always "perfect and complete."

axioma	axiom
dignitas	axiom

In *Comm. Post. An. l.* 1 *t.* 14 and the index to *Opera Logica*, Zabarella says these two mean the same thing; the first term is the Greek, the second the Latin.

principium   beginning (as of a chapter)
         beginning-principle (as a premise in a syllogism)

Translators conventionally choose "beginning" or "principle" as the context suggests. But this masks the role in Zabarella of a principle as an ontological beginning. If this is not constantly borne in mind (especially in *l.* 2), it is often difficult to appreciate Zabarella's point or even just the meaning of some sentences. So, for *principium*, "beginning" for, say, the beginning of a chapter, is here used, but "beginning-principle" when "principle" would be a more traditional translation. The strategy parallels the emerging preference for the hyphenated "starting-point" among translators of *principium*'s Greek equivalent, αρχη (*archē*).

in (eo) quod quid est   because of what (it) is

This translates the Greek ἐν τῷ τί ἐστι.

proportione   in comparative relation

The term should probably be considered against the background of medieval theories of the analogy of proportion and proportionality.

temperies   temper
temperatura   balance

These are synonymous technical terms in medieval and Renaissance medicine referring to a balance of elements, humors, or properties. To maintain the difference, two English words are used, but Zabarella may or may not have thought there was any substantive difference.

species   species
genus   genus, kind

differentia   differentia, difference

In Zabarella's time the English phrase was "genus and difference" not "genus and differentia." John Locke used "differentia" in 1691; he did not need to explain the term, but he did ask his reader's indulgence for using it. Here, "genus" and "differentia" are used when the context involves the relationship between genus and species; "kind" and "difference" otherwise. But the distinction is not always sharp and would not have been a strong one for Zabarella and his readers.

discrimen   discriminating difference
distinctio   distinction

For Zabarella, generally, *discrimen* is ontological, *distinctio* is epistemological. A *discrimen* exists in the world; a *distinctio* is made by us. A *discrimen* is the difference by which one genus differs from another. A *distinctio* is the product of a cognitive act, the act of distinguishing, the act of recognizing a *discrimen*. When we distinguish, we make a distinction; it is we who distinguish A and B. It is not the difference that distinguishes them.

accidentia   accident
affectio   affection

In his commentary on *Posterior Analytics* l. 1 t. 57, Zabarella says that for Aristotle, "accident" and "affection" mean the same thing and says that if there is any useful distinction (as Averroës held) it would be that affections are accidents more known. But whether formally there is any difference, Zabarella consistently uses *affectio* when presenting an accident as a term in the conclusion of a syllogism in which a beginning-principle and an underlying subject are the other two terms. In the works here, Zabarella uses *accidentia* very seldom.

| *prima materia* | first matter |
| *prima motor* | first mover |

The conventional sixteenth- and early seventeenth-century English renderings are used. None of the terms were typically capitalized, and none are here. Only later did "prime matter" and "prime mover" gain currency.

| *immobilis motor* | immobile mover |

The Latin *immobilis* translates Aristotle's *akinēton*. Both the Greek and Latin can mean "immovable," "motionless," or "unmoved." Dictionaries in Zabarella's time always gave the first and sometimes the second, but many commentators then and now have preferred the third. The ambiguity is here retained.

philosophicus	philosopher
divinus philosophicus	philosopher of divinity
primus philosophicus	philosopher of metaphysics
metaphysicus	metaphysician, metaphysical
Metaphysicos	the *Metaphysics*

A *metaphysicus*, "metaphysician," is presumably the same as a *primus philosophicus*, "first philosopher," "philosopher of first philosophy," or here "philosopher of metaphysics," but a distinction is retained nonetheless. Zabarella does not use *theologia* or *theologicus*.

| *scriptor* | writer |
| *author* | author |

The meanings of *auctor* and *author*, what in earlier times were simply alternate spellings of one word, were distinguished in Renaissance dictionaries. The first suggested an originator, the second an authority, and Zabarella consistently uses the second. This is unsurprising given that he believes scientific learning should be

done largely by reading and studying established texts, and he is usually discussing such texts. The sense of authority should always be read into his term.

habitus   habit

A habit (*habitus*) is, generally, something one has (*habet*) in mind, and only specifically a mental disposition or customary practice. For Zabarella, knowledge one has of a certain subject matter is a habit. The ambiguity between the general and the specific meanings is central to Zabarella's argument in *l.* 1 *c.* 2–3.

anima   soul
animus   man's soul, our soul, my soul, someone's soul, etc.
mens   mind

Feminine *anima* translates Greek *psyche*, as in the title of Aristotle's book *De Anima* (*On the Soul*). The word refers to the soul in any animate being, the bundle of vital capacities that allow it to function as a living being. Masculine *animus* is the rational soul, the part or function of *anima* in man that thinks. In some phrases, Zabarella uses *animus* and *mens* interchangeably, but in others he sticks strictly to one or the other. For example, he frequently explains that, after it has obtained the conception it seeks, our soul (*animus*) rests (*acquiescit, conquiescit,* or *quiescit*). In Locke and Hobbes, it is the mind, not the soul, that rests. This suggests "mind" as a translation of *animus*. But there is surely an indication here of a substantive difference in psychological ontology between the late Renaissance and the early modern, so "mind" is here reserved for *mens*.

Notes to the Text

1. dum] eam d E
2. causa] in causa A
3. veram] verum A
4. significatio] significato B
5. sint] sunt E
6. externa] extrema A
7. discipulis] disciplinis E
8. universè *et passim*] universim A
9. sumpto] accepto A
10. dicamur] dicantur A
11. explicari] explicare B
12. imitetur] imitatur A
13. dicamus] dicantur A] dicamur B
14. *B, C, d, and E lack* Aristoteles
15. *B, C, d, and E lack* quo Aristoteles sensus ipsos disposuit
16. *A lacks* quoque
17. *C, d, and E lack* à
18. quiddam] quoddam E
19. *C, d, and E lack* à
20. tantam] tantum A
21. ut] et d E
22. *B, C, d, and E lack* quidem
23. saepe] ut plurimum A
24. his] his autem A
25. cuiusque] utriusque A
26. possunt] possent A B
27. aliter] alter B d E
28. *B, C, d, and E lack* Aristoteles
29. cui] uti A
30. *B lacks* à
31. *B, C, d, and E lack* quidem
32. considerata] consideratas B
33. *The only instance of* universim *in* De Methodis *and* De Regressu *not changed to* universè *after the first edition.*
34. mutationem] tractationem A
35. unam] una B
36. animatum] animantium d E
37. direxerit] direxit B
38. *A lacks* quod etiam ab Avicenna notatur in prooemio sexti Naturalium
39. nullam] nullum B
40. Quod] Quòd B
41. posset] possit A B
42. alias] aliàs B
43. queant] valeant A

BOOK II

1. simul] quidem *E*
2. *B, C, d, and E lack* autem
3. sit] fit *A B C*
4. ullum] ultimum *C d E*
5. tam] tum *A*
6. *B, C, d, and E lack* scientiae illius
7. aliqui] aliquà *A*
8. contradictio] contradictis *A*
9. inanima] inanimata *E*
10. definitione] definitiones *B*
11. coeperamus] ceperamus *A*
12. possit] valeat eum *A*
13. aptae sint] valeant *A*
14. quod] et quod *B*
15. propriamne] propriumne *A*
16. qua] quo *A*
17. habentem] habens *A*
18. *C, d, and E lack* autem
19. deducuntur] derivantur *A*
20. possimus] valeamus *A*
21. facile] facilè *B*
22. ab] ob *A*
23. verum] veru *A*
24. scientiam] scientia *B*
25. potiti] potiri *C d E*
26. quia] quà *A*
27. *B, C, d, and E lack* à fine moveri, aliud sit
28. à fine moveantur] fine moveamur *B C d E*
29. possit] valeat *A*
30. quia] quae, quia *A*
31. *A lacks* est

32. appellari] appellaria *A*
33. accipit] accepit *A*
34. moralem] naturalem *A*
35. sed] sed potius *A*
36. *A lacks* certè
37. *C, d, and E lack* haec
38. ex] et *A*
39. *One of only two instances of* valere *in* De Methodis *and* De Regressu *not changed to something else after the first edition.*
40. ne] nec *A*
41. possimus] valeamus *A*
42. *C, d, and E lack* et
43. tota] nota *d E*
44. possunt] valent *A*
45. illas] illius *A*
46. praemittit] praemitti *d E*
47. *Zabarella here uses a Greek genitive.*
48. strictius] distinctius *C d E*
49. ob] ad *A*
50. possimus] valeamus *A*
51. apta sint] valeant *A*
52. *B, C, d, and E lack* à
53. possimus] possumus *A*
54. *B, C, d, and E lack the paragraph* Praeterea non tres . . . auferri.
55. *A lacks* quidem
56. *B lacks* pars
57. praecipua] praecipuè *C d E*
58. *B, C, d, and E lack* quidem

59. dispositione] disputatione
    d E

60. conferant] conferat *A B*

61. lucem] lumen *A*

62. *B, C, d, and E lack* quidem

63. illa] ea *E*

64. dicimus] elicimus *C d E*

65. posset] potest *A*

66. vel] ut *C d E*

Note on References

⁂

For everything in *Opera Logica*, Zabarella's point of reference is the corpus (or what was thought to be the corpus) of Aristotle. Zabarella's *Commentary on the Two Books of the Posterior Analytics* itself is three-quarters as big as everything else in the *Opera* combined, and the other parts, including *On Methods* and *On Regressus*, are largely commentaries on specialized topics in scholastic Aristotelian philosophy.

In several ways, the Aristotelian corpus as Zabarella knew it differs from the corpus as we know it. One difference is that in Zabarella's community of colleagues and readers, the corpus circulated with Averroës' commentaries interleaved with blocks of Aristotle's writing. In many works, each block, called a *contextus*, had a number, here translated as "text no." and abbreviated *t*. For the works that had these numbers, they were the standard mechanism by which passages were cited.

To make it easier to correlate Zabarella's comments with other parts of his writings, with the writings of other commentators, and with published Renaissance printings, citations to Aristotle and to Averroës are to the divisions as Zabarella knew them. The modern Bekker, book, and chapter numbers that correspond to those sections are then given in parentheses. The citation does not attempt to identify, narrowly, the passage Zabarella is referring to, but rather identifies the full extent of the Renaissance unit in which the cited passage lies. Hence, "*Posterior Analytics l.* 2 *t.* 36–47 (93a1–19 = *l.* 2 *c.* 8–10)" indicates that the passage Zabarella cites and knew as *liber* 2, *contextus* 36 through 47, corresponds to Bekker 93a1–19 and that this precisely matches Bekker Book 2, chapters 8–10. "*On Interpretation c.* 1–4 (16a1–17a37 = *c.* 1–6)" indicates that what Zabarella knew as chapters 1 through 4, we know as 1

through 6. "*On the Soul l.* 1 *t.* 3 (402a7–10 in 1.1)" indicates that *liber* 1, *contextus* 3 corresponds to a passage, Bekker 402a7–10, that lies within what we know as Book 1, chapter 1. For the *Posterior Analytics*, boundaries in Zabarella's own edition (*Comm. Post. An.*) are presumed authoritative; for other titles, AOAC is used. For citations to Plato, the parenthetical reference is to Stephanus pagination. For citations to Galen, it is to Kühn pagination; for Avicenna, the numbering in Bakhtiar.

Volume 1 of *AOAC* has three parts. Though *AOAC*'s table of contents does not indicate this, the numbering of folios restarts partway through each of the three. The first numbered sequence in part 1 is here cited as *pt.* 1a, the second as *pt.* 1b; the first sequence in part 2 as *pt.* 2a and the second as *pt.* 2b. The second pair matches the labels used in the binding of the 1962 reprint by Minerva.

Though citations here are to *AOAC*, Zabarella's quotations of Aristotle and Averroës are generally not taken from the translations in *AOAC* or from those in other common Renaissance editions. The passages of Galen are not those by Niccolò Leoniceno. Thus, translations from Greek are possibly Zabarella's own. His Latin translations of Averroës would have had to come from a Hebrew edition. Some clues to what editions Zabarella was using are given in the notes.

Some writings that Zabarella and his contemporaries presumed were Aristotle's own we now believe to be otherwise. Zabarella presumes that titles and the order in the printed corpora were Aristotle's choice, and some of Zabarella's arguments rely on this.

The reader should be cautious about presuming that what Zabarella claims is stated in a cited passage is as unambiguously stated as Zabarella suggests.

Works with no author listed are Aristotle's or Zabarella's. Thus, "*Topics*" is Aristotle's; "Cicero, *Topics*" is Cicero's.

References for the Greek commentators are to *CAG*, taken from Schicker. These references can be used to find the relevant passages in the English translations in the series edited by Richard Sorabji, *Ancient Commentators on Aristotle* (London, 1987–).

ABBREVIATIONS

AOAC	Aristotle, Averröes, et al. *Aristotelis opera cum Averrois commentariis*. Venice: Iunctae, 1562. Reprint, Frankfurt am Main: 1962.
Bakhtiar	Laleh Bakhtiar. *The Canon of Medicine*. Great Books of The Islamic World, 1999.
CAG	*Commentaria in Aristotelem Graeca*. 23 vols. Berlin: Reimer, 1882–1909.
Comm. Post. An.	Jacopo Zabarella. *Commentarii in libros duos Posteriorum Analyticorum*. Venice: Meietus, 1582.
Galen	*Opera omnia*. Edited by Karl Gottlob Kühn. 22 vols. Leipzig: Knobloch, 1821–33.
Holyoake	Francis Holyoake. *Dictionarium Etymologicum Latinum*. Oxford: William Turner, 1627.
Meth.	Jacopo Zabarella. *On Methods*.
Regr.	Jacopo Zabarella. *On Regressus*.
Schicker	Jacopo Zabarella. *Über die Methoden = De methodis; Über den Rückgang = De regressu*. Translated with an introduction and commentary by Rudolf Schicker. Munich: Wilhelm Fink, 1995.

c.	*caput, capitulum*, chapter
d.	*doctrina*
fn.	*fen* (a subdivision in Avicenna, *Liber Canonis*)
fol.	folio
l.	*liber*, book
ln.	line
lo.	*locus*, locus
p.	page

par.	paragraph
pt.	part
q.	*quaesitum,* question
s.	*sectio,* section
sg.	segment of a page in *AOAC*
su.	*summa*
t.	*contextus,* text no.
tr.	*tractatus,* tract
v.	*volumen,* volume

Notes to the Translation

BOOK I

1. Galen, *The Art of Medicine* (1:305).

2. Capitalization here follows the source, but Zabarella is not consistent. The capitalized word has the broader sense, but the uncapitalized one can also.

3. See Note on the Text and Translation, s.v. *habitus*.

4. Zabarella, *De natura logicae* (the first work in *Opera Logica*), *l.* 1 *c.* 5.

5. *On Interpretation l.* 1 *c.* 4 (16b26–17a37 = *c.* 4–6). Zabarella's translation puts more stress on the final clause than is in the Greek or in other Latin translations. A closer rendering would be "but (as was said) by convention." Other Latin translations were *sed* (*quemadmodum dictum est*) *ex instituto* or small variants thereof. The Greek is ἀλλ' ὥσπερ εἴρηται κατὰ συνθήκην (17a1–2).

6. Above all in Plato's *Cratylus*, but there are also key passages in the *Phaedrus*, *Philebus*, *Sophist*, and *Statesman*.

7. Can be reciprocated: the two have the same extent. It was said that a whole and its parts could be reciprocated, as could a definition and the thing being defined.

8. *Posterior Analytics l.* 2 *t.* 42–46 (93b15–94a14 in *l.* 2 *c.* 8–9). See Zabarella's commentary on this passage, esp. *t.* 43, in *Comm. Post. An.* See also *Meth. l.* 3 *c.* 9, and *Regr. c.* 4 *par.* 14.

9. For this and the next few paragraphs, see Note on the Text and Translation, s.v. *habitus*.

10. Galen, *The Art of Medicine* (1:305). These are the works first words, though not in the translation by Leoniceno.

11. *Posterior Analytics l.* 1 *t.* 1 (71a1–11 in *l.* 1 *c.* 1).

12. On the imagery of "are led," see Note on the Text and Translation, s.v. *via*.

13. For the demonstration of first matter from change of substance (generation), see *Regr. c.* 4 and 5. For the demonstration of the unmoved mover from eternal motion and vice versa, see *Regr. c.* 6.

14. On "discriminating difference" and the act of distinguishing, see Note on the Text and Translation, s.v. *discrimen*.

15. *Physics l.* 1 *t.* 1–5 (184a10–b14 = *l.* 1 *c.* 1).

16. Zabarella presumes the reader knows that compositive order is the disposition that begins with causes and beginning-principles and proceeds to effects. Resolutive order proceeds in the opposite direction.

17. *Metaphysics l.* 7 *t.* 23 (1032a28–b31 in *l.* 7 *c.* 7). Aristotle describes and exemplifies two procedures, but he does not call them resolution (*analysis*) and composition (*synthesis*), as Zabarella frequently implies. Aristotle says the process that proceeds from the beginning-principle and the form is called *noēsis* (*intellectio* in AOAC). That which proceeds from the end of the thinking is called *poiēsis* (*effectio* in AOAC). A marginal gloss added by the editor in AOAC reads, "*Hic habes methodos Resolutivam ac Compositivam*" (Here you have resolutive and compositive methods).

18. *Nicomachean Ethics l.* 7 *c.* 8 (1150b29–1151a28 = *l.* 7 *c.* 8).

19. Ibid. *l.* 1 *c.* 13 (1102a5–1103a10 = *l.* 1 *c.* 13). For Aristotle's *kat' aretēn* (1102a5), Zabarella uses *ex virtute* (out of virtue), instead of the conventional *secundùm virtutem* (in accordance with virtue).

20. *Metaphysics l.* 7 *t.* 23 (1032a28–b31 in *l.* 7 *c.* 7).

21. *On the Parts of Animals l.* 2 *c.* 1 (646a8–647b9 = *l.* 2 *c.* 1).

22. *Physics l.* 1 *t.* 1–5 (184a10–b14 = *l.* 1 *c.* 1). Zabarella frequently refers to the beginning of the *Physics*, what we know as *c.* 1 and what is *l.* 1 *su.* 1 in AOAC, as a proem. It may have been titled such in the edition he used.

23. *Posterior Analytics l.* 1 *t.* 166 (85b28–86a3 in *l.* 1 *c.* 24); *l.* 2 *t.* 1 (89b23–31 in *l.* 1 *c.* 1).

24. *Physics l.* 1 *t.* 1–5 (184a10–b14 = *l.* 1 *c.* 1).

25. *Metaphysics l.* 5 *t.* 1 (1012b34–1013a23 = *l.* 5 *c.* 1).

26. *Nicomachean Ethics l.* 1 *c.* 4 (1095a14–b13 = *l.* 1 *c.* 4).

27. Ibid.

28. Ibid. *l.* 1 *c.* 3 (1094b11–1095a13 = *l.* 1 *c.* 3).

29. See the comparison of "relative" and "absolute" in *Meth. l.* 2 *c.* 3 *par.* 5.

30. *On the Soul l.* 2 *t.* 66–120 (418a26–424a16 = *l.* 2 *c.* 7–11).

31. *On the Soul.* Taste: *l.* 2 *t.* 101–5 (421a8–422b16 = *l.* 2 *c.* 10); touch: *l.* 2 *t.* 106–20 (422b17–424a16 = *l.* 2 *c.* 11).

32. *On the History of Animals l.* 1 *c.* 6 (490b7–491a27 = *l.* 1 *c.* 6).

33. *On the Parts of Animals l.* 2 *c.* 10 (655b28–657a11 = *l.* 2 *c.* 10).

34. *On Length and Shortness of Life c.* 4 (467a6–b9 = *c.* 6).

35. *Physics l.* 1 *t.* 57–70 (189b30–191a22 in *l.* 1 *c.* 6–7), especially *t.* 62 (190a31–b5). See *De Regressu c.* 4.

36. Averroës, long commentary on *Physics l.* 1 *t.* 57 (189b30–32 in *l.* 1 *c.* 6–7), *AOAC v.* 4 *fol.* 34 *sg.* G–L.

37. Zabarella is drawing a contrast not between the use of two mutually exclusive procedures but between the roles played by the sequence (i.e., the order) of a presentation's components and whether the components are connected by demonstrative inference (i.e., whether method is employed).

38. *Metaphysics l.* 5 *t.* 1 (1012b34–1013a23 = *l.* 5 *c.* 1).

39. The phrase "branch of learning" translates *facultas* (*facultatis,* genitive) and presumes Zabarella does not mean faculties of the soul or mind.

40. See Note on the Text and Translation, *s.v. author.*

41. *Physics l.* 1 *t.* 1 (184a10–16 in *l.* 1 *c.* 1).

42. *Metaphysics l.* 5 *t.* 1 (1012b34–1013a23 = *l.* 5 *c.* 1).

43. Averroës, middle commentary on *On Meteorology l.* 1 *su.* 1 *c.* 1–5 (338a20–341a36 = *l.* 1 *c.* 1–3), *AOAC v.* 5 *fol.* 403 *sg.* G —*fol.* 406 *sg.* I.

44. A fallacy Aristotle describes in *Sophistical Refutations c.* 5 166b38–167a21. Zabarella cites it again in *Regr. c.* 4 *par.* 4.

45. *Physics l.* 1 *t.* 1–5 (184a10–b14 = *l.* 1 *c.* 1).

46. Averroës, proem to long commentary on the *Physics, AOAC v.* 4 *fol.* 1 *sg.* A —*fol.* 5 *sg.* F.

47. Averroës, proem to long commentary on *Posterior Analytics*, *AOAC v.* 1 *pt.* 2a *fol.* 1 *sg.* A—*fol.* 9 *sg.* E.

48. *Posterior Analytics l.* 2 *t.* 48 (94a20–24 in *l.* 2 *c.* 11).

49. Predication in which the predicate is in the nominative case, as when a genus is predicated of a species (man is animal, *homo est animal*) or an attribute of substance (lettuce is cold, *lactuca est frigida*).

50. *On the Heavens l.* 1 *t.* 92 (277b29–278a22 in *l.* 1 *c.* 9). Aristotle's Greek has definite articles: τὸ μὲν εἶδος, τὸ δ' εἶδος ἐν τῇ ὕλῃ.

51. The study of solids.

52. In this argument, Zabarella is presuming that the order of the books as he had them is the order that Aristotle intended.

53. That is, from the middle of the earth and so also, in Aristotelian cosmology, from the middle of the universe.

54. Averroës, long commentary on *Posterior Analytics l.* 1 *t.* 2 (71a11–17 in *l.* 1 *c.* 1), *AOAC v.* 1 *pt.* 2a *fol.* 18 *sg.* C—*fol.* 20 *sg.* D.

55. Such a "consideration of the mind" (*mentis consideratio*) plays a key role in a regressus. See *Regr. c.* 5.

56. Through the rest of this chapter, much of Zabarella's argument relies on his assumption, sometimes warranted, sometimes not, that the sequence of Aristotle's books as published was the sequence Aristotle intended.

57. The nutritive soul, commonly called by others the vegetative soul and so called by Zabarella in the next paragraph, has three faculties: nutritive, augmentative, and generative.

58. The sixth book of his *Naturalia* is Avicenna's *Liber de anima*.

59. See *On the Soul l.* 2 *t.* 6–7 (412a26–b9 in *l.* 2 *c.* 1).

60. *On Sense and Sensibles l.* 1 *c.* 1 (436a1–439a6 = *l.* 5 *c.* 1).

61. Ibid. *l.* 1 *c.* 1 (436a1–439a6 = *l.* 1 *c.* 1).

62. *On Health and Disease* is known only in a sixty-eight-word fragment and is considered spurious. The fragment appeared in *AOAC v.* 6 *pt.* 2 *fol.* 159 *sg.* D–F.

63. *On Sense and Sensibles* l. 1 c. 1 (436a1–439a6 = l. 1 c. 1).

64. Zabarella's statement might suggest that he was using an edition that treated *On Life and Death* and *On Respiration* as one unit separate from *On Youth and Old Age*.

BOOK II

1. Galen, *The Art of Medicine* (1:305).

2. *Physics* l. 1 t. 1–5 (184a10–b14 = l. 1 c. 1).

3. *On Meteorology* l. 1 su. 1 c. 1 (338a20–339a9≈ l. 1 c. 1).

4. *Nicomachean Ethics* l. 1 c. 4 (1095a14–b13 = l. 1 c. 4). At 1095a31–32, Aristotle says there is a difference between arguments "from beginning-principles (*apo tōn archōv*)" and those "to beginning-principles (*epi tas archas*)." He repeats the distinction a few lines later.

5. *Posterior Analytics* l. 1 t. 179 (87a39–b4 = l. 1 c. 28).

6. See the comparison of relative and absolute in *Meth.* l. 1 c. 7 *par.* 6.

7. Plato, *Phaedrus* (277b5–8).

8. *Nicomachean Ethics* l. 1 c. 7 (1097a15–1098b8 = l. 1 c. 7).

9. *On Interpretation* l. 1 c. 1–4 (16a1–17a37 = c. 1–6).

10. *Prior Analytics* l. 1 s. 1 c. 1 (24a10–b30 = l. 1 c. 1).

11. *Posterior Analytics* l. 1 t. 8 (71b15–19 in l. 1 c. 2).

12. *Physics* l. 2 t. 3 (192b20–23 in l. 2 c. 1).

13. *Metaphysics* l. 7 t. 23 (1032a28–b31 in l. 7 c. 7).

14. Galen, *The Art of Medicine* (1:307).

15. "to make the reader more teachable" (*ad lectoris docilitatem*). In classical rhetoric, one task of an introduction is *captatio benevolentiae*, to make the reader or listener attentive, benevolent, and teachable (*docilis*).

16. *On Meteorology* l. 1 su. 1 c. 1 (338a20–339a9 ≈ l. 1 c. 1).

17. *Physics* l. 1 t. 1–5 (184a10–b14 = l. 1 c. 1).

18. Avicenna, *Liber Canonis*. See note to *Meth.* l. 2 c. 14 *par.* 12 below.

19. *On the Heavens* l. 1 t. 7 (268b26–269a2 in l. 1 c. 2).

20. *Physics l.* 1 *t.* 1–5 (184a10–b14 = *l.* 1 *c.* 1).

21. Zabarella, *De natura logicae* (the first work in *Opera Logica*) *l.* 1 *c.* 2.

22. *Physics l.* 1 *t.* 1 (184a10–16 in *l.* 1 *c.* 1).

23. The syllogism Zabarella sees in *Physics l.* 1 *t.* 1 (184a10–16 in *l.* 1 *c.* 1) is the following. Major premise: Knowledge—perfect scientific knowledge—of certain things cannot be had except from known first beginning-principles. Minor premise: Natural things are such things. Conclusion: Perfect scientific knowledge of natural things cannot be had except from known first beginning-principles.

24. *Posterior Analytics l.* 1 *t.* 7 (71b9–15 in *l.* 1 *c.* 2).

25. Zabarella is referring to the first sentence of the *Physics*, in which Aristotle uses the Greek *methodous* (pl. of *methodos*). Zabarella is staying unusually close to the Greek here in using the Latin *methodos*. Translators typically used *doctrinas* or *scientias*, as Zabarella himself does in his extended comments on the passage in *Comm. Post. An. l.* 1 *c.* 1. See those comments for more on Zabarella's reasoning here.

26. *Metaphysics l.* 7 *t.* 23 (1032a28–b31 in *l.* 7 *c.* 7).

27. Zabarella has returned to discussing *Physics l.* 1 *t.* 1. Aristotle does not use the word (*methodos*) in *Metaphysics l.* 7 *t.* 23.

28. In this paragraph, "intention" is the medieval technical term for "idea" or "concept."

29. *Physics l.* 2 *t.* 23 (194a27–33 in *l.* 2 *c.* 2).

30. Ibid. *l.* 1 *t.* 1–5 (184a10–b14 = *l.* 1 *c.* 1).

31. In *AOAC, On the Heavens* (the beginning of *AOAC* volume 5) does immediately follow the *Physics* (all of *AOAC* volume 4). Though the ordering is still conventional, we cannot assume Aristotle intended it.

32. Averroës, preface to *Physics l.* 1, *AOAC v.* 4 *fol.* 1 *sg.* C.

33. *Physics l.* 1 *t.* 1–5 (184a10–b14 = *l.* 1 *c.* 1).

34. Recall from *Meth. l.* 1 *c.* 2 *par.* 1 that "method" can be used in a wide sense to include both method in the narrower sense and order.

35. *Metaphysics l.* 7 *t.* 23 (1032a28–b31 in *l.* 7 *c.* 7).

36. *Nicomachean Ethics l.* 3 *c.* 3 (1112a18–1113a14 = *l.* 3 *c.* 3).

37. Ibid. *l.* 1 *c.* 4 (1095a14–b13 = *l.* 1 *c.* 4).

38. *Metaphysics l.* 7 *t.* 23 (1032a28–b31 in *l.* 7 *c.* 7).

39. *Nicomachean Ethics l.* 6 *c.* 5 (1140a24–b30 = *l.* 6 *c.* 5); *l.* 7 *c.* 8 (1150b29–1151a28 = *l.* 7 *c.* 8).

40. Ibid. *l.* 1 *c.* 7 then *c.* 9 (1097a15–1098b8 = *l.* 1 *c.* 7 then 1099b9–1100a9 = *c.* 9)

41. Galen, *On the Constitution of the Medical Art* (1:224–304). In published works of Galen in the sixteenth century, chapter numbers, when they existed at all, were not standard. If Zabarella is referring to what Galen says at Kühn 1:227, 1:295, and 1:302, then Zabarella is not referring to chapters as numbered in the *Opera* of Venice: Junctas, 1565, or Basil: 1549. (Basil: 1531 and Basil: 1571 do not have chapter numbers.) Schicker has (the German for) "first and penultimate" instead of "first, last, and penultimate" and cites Kühn 1:226 and 1:302.

42. Avicenna, *Liber Canonis*. See note to *Meth. l.* 2 *c.* 13 *par.* 1 below.

43. Averroës, *Colliget*. See note to *Meth. l.* 2 *c.* 12 *par.* 1 below.

44. See Note on the Text and Translation s.v. *temperies*.

45. Averroës, *Colliget*. See note to *Meth. l.* 2 *c.* 12 *par.* 1 below.

46. Galen, *On the Constitution of the Medical Art* (1:257–60 = *c.* 10).

47. Zabarella, *De natura logicae* (the first work in *Opera Logica*) *l.* 2 *c.* 4.

48. *Physics l.* 2 *t.* 26 (194b8–15 in *l.* 2 *c.* 2).

49. *Nicomachean Ethics l.* 1 *c.* 13 (1102a5–1103a10 = *l.* 1 *c.* 13).

50. Galen, *On the Constitution of the Medical Art* (1:257–60 = *c.* 10). Schicker has "chapter 2" instead of "chapter 11." Schicker was apparently unaware that the chapter numbers Zabarella had for Galen were different than those in Kühn. See, e.g., *Meth. l.* 2 *c.* 14 *par.* 3 below. So when he could not see in (Kühn's) chapter 11 what Zabarella says was there, Schicker figured that what looked like an Arabic numeral 11 was actually a Roman numeral 2. But this would be exceptional and at least one edition (though admittedly posthumous), the 1608, has "undecimo" instead of a numeral. And although the subject is not treated in the same depth

in the chapter we know Zabarella considered a summary (his *c.* 11, Kühn's *c.* 10), as in the passage Schicker cites (Kühn's *c.* 2), we should assume that an "11" was intended and that Kühn's *c.* 10 is the chapter Zabarella was referring to.

51. *Posterior Analytics l.* 1 *t.* 2 (71a11–17 in *l.* 1 *c.* 1).

52. *Metaphysics l.* 7 *t.* 23 (1032a28–b31 in *l.* 7 *c.* 7).

53. Zabarella, *De natura logicae* (the first work in *Opera Logica*) *l.* 2 *c.* 4.

54. Avicenna, *Liber Canonis l.* 1 *fn.* 1 *d.* 1 *c.* 2 (*v.* 1 *s.* 12–24).

55. Averroës, *Colliget l.* 2 *c.* 1, *AOAC v.* 10 *fol.* 12 *sg.* I—*fol.* 14 *sg.* D. See note to *Meth. l.* 2 *c.* 12 *par.* 1 below.

56. Averroës, *Colliget.* This work was printed in volume 10 (also known as Supplement 1) of *AOAC.* It is composed of seven books: *l.* 1 *Anatomy, l.* 2 *Health, l.* 3 *Sicknesses and accidents, l.* 4 *Signs of health and sicknesses, l.* 5 *Nutriments and medicines, l.* 6 *Health regimen, l.* 7 *Cure for sickness, or the character of health.*

57. Ibid. *l.* 1 *c.* 1, *AOAC v.* 10 *fol.* 3 *sg.* D—*fol.* 4 *sg.* D.

58. Ibid.

59. Ibid.

60. See Note on the Text and Translation, s.v. *temperies.*

61. Averroës, *Colliget l.* 1 *c.* 1, *AOAC v.* 10 *fol.* 3 *sg.* D—*fol.* 4 *sg.* D.

62. Ibid. *c.* 2 (in the proem before *v.* 1), *AOAC v.* 10 *fol.* 1 *sg.* H—*fol.* 3 *sg.* C.

63. Ibid. *l.* 2 *c.* 1, *AOAC v.* 10 *fol.* 12 *sg.* I—*fol.* 14 *sg.* C.

64. For "complexional form," see, e.g., Averroës, *Colliget l.* 2 *c.* 1, *AOAC v.* 10 *fol.* 12 *sg.* I—*fol.* 14 *sg.* C; Avicenna, *Liber Canonis l.* 1 *fn.* 1 *d.* 1 *c.* 2 (*v.* 1 *s.* 12–24, esp. 15); and John Marebon, *Later Medieval Philosophy* (London: Routledge, 1987), p. 103.

65. Averroës, *Colliget l.* 1 *c.* 1, *AOAC v.* 10 *fol.* 3 *sg.* D—*fol.* 4 *sg.* D.

66. Ibid. *l.* 3 *c.* 1, *AOAC v.* 10 *fol.* 33 *sg.* A—H.

67. Ibid. *l.* 4, *AOAC v.* 10 *fol.* 58 *sg.* K—*fol.* 86 *sg.* K.

68. Ibid. *l.* 5, *AOAC v.* 10 *fol.* 86 *sg.* L—*fol.* 132 *sg.* M.

69. Ibid. *l. 5 c.* 30, *AOAC v.* 10 *fol.* 99 *sg.* F—*fol.* 100 *sg.* I; *c.* 31, *AOAC v.* 10 *fol.* 100 *sg.* K—*fol.* 101 *sg.* G.

70. *Posterior Analytics l.* 1 *c.* 2 (71b9–72b4 = *l.* 1 *c.* 2).

71. Ibid. *l.* 1 *c.* 4 (73a21–74a4 = *l.* 1 *c.* 4).

72. Averroës, *Colliget l.* 6, *AOAC v.* 10 *fol.* 133 *sg.* A—*fol.* 142 *sg.* M; *l.* 7, *AOAC v.* 10 *fol.* 143 *sg.* A—*fol.* 172 *sg.* L.

73. Avicenna, *Liber Canonis*. Latin editions were printed many times from the 1470s through the end of the sixteenth century. An Arabic edition was published in Rome in 1593. The work is composed of five books (here in a close English rendering of the vulgate Latin translation by Gerard of Cremona): *l.* 1 *Things universal in the science of medicine, l.* 2 *Simple medicines, l.* 3 *Particular manifest and hidden sicknesses that occur in members of the body from head to feet, l.* 4 *Particular sicknesses that do not occur in one member, and cosmetics, l.* 5 *Compounding of medicines and their antidotes.* The first book has four parts, each called a *fen: fn.* 1 *The definition of medicine and its subject and natural things, fn.* 2 *The division of sicknesses by causes and universal accidents, fn.* 3 *Conservation of health, fn.* 4 *The division of ways of medicating according to universal sicknesses.*

74. See Note on the Text and Translation, *s.v. temperies.*

75. Cf. Andreas Vesalius, *De humani corporis fabrica* (Basel: Oporinus, 1543).

76. *Metaphysics l.* 7 *t.* 23 (1032a28–b31 in *l.* 7 *c.* 7). See note to *Meth. l.* 1 *c.* 6 *par.* 3.

77. Knowledge of causes is gotten from the natural philosopher.

78. Galen, *Art of Medicine* (1:365).

79. Galen, *On the Constitution of the Medical Art* (1:224–304). See note to *Meth. l.* 2 *c.* 10 *par.* 4.

80. Ibid. (1:257–60 = *c.* 10). Schicker glossed "in this chapter" with "(the tenth!)." But in Venice: Junctas, 1565, the chapter is indeed numbered 11 and labeled a summary (*epilogus*). So it is here assumed that the edition Zabarella was using had chapter numbers matching those in Venice: Junctas, 1565, at least in this section. These chapter numbers have been used to cross-reference into Kühn.

81. Ibid. (1:254–57 = c. 9).

82. Ibid. (1:245–54 = c. 7).

83. Ibid. (1:260–65 in c. 11–12).

84. Galen, *The Order of My Books* (19:49–61).

85. Galen, *The Best Doctor is also a Philosopher* (1:53–63).

86. Galen, *The Art of Medicine* (1:309).

87. Ibid. (1:306).

88. Schicker cites Galen, *On the Constitution of the Medical Art* (1:226).

89. Galen, *The Art of Medicine* (1:307).

90. Ps.-Galen, *Introduction, or The Physician* (14:688), s.v. Herophilus, the anatomist who was born in Chalcedon and practiced in Alexandria in the early third century BCE.

91. Galen, *The Art of Medicine* (1:307). Galen's term in the passage is *epistemē*.

92. Avicenna, *Liber Canonis* l. 1 *fn.* 1 d. 1 c. 1 (*v.* 1 s. 7–11). See note to *Meth.* l. 2 c. 13 *par.* 1 above. The first words of the main body of the *Canon*, after Avicenna's preface, are these: "Medicine is the science by which we learn the various states of the human body in health and when not in health, and the means by which health is likely to be lost and, when lost, is likely to be restored to health. In other words, it is the art whereby health is conserved and the art whereby it is restored after being lost" (Bakhtiar translation).

93. Averroës, *Colliget* l. 1 c. 1, AOAC v. 10 fol. 3 sg. D—fol. 4 sg. D.

94. See note to *Meth.* l. 1 c. 14 *par.* 22, s.v. the fragment *On Health and Disease*.

95. Averroës, *Colliget* l. 1 c. 1, AOAC v. 10 fol. 3 sg. D—fol. 4 sg. D.

96. Galen, *The Art of Medicine* (1:307).

97. *Physics* l. 1 t. 1–5 (184a10–b14 = l. 1 c. 1).

98. *Meteorology* l. 1 su. 1 c. 1 (338a20–339a9 ≈ l. 1 c. 1).

99. *Metaphysics* l. 7 t. 23 (1032a28–b31 in l. 7 c. 7). See note to *Meth.* l. 1 c. 6 *par.* 3.

100. *Nicomachean Ethics* l. 6 c. 5 (1140a24–b30 = 6.5); l. 7 c. 8 (1150b29–1151a28 = l. 7 c. 8).

101. Ibid. l. 1 c. 4 (1095a14–b13 = l. 1 c. 4).

102. Ibid. l. 6 c. 5 (1140a24–b30 = l. 6 c. 5); l. 7 c. 8 (1150b29–1151a28 = l. 7 c. 8).

103. See *Physics* (195a7–11 in l. 2 c. 3). "Some things cause each other reciprocally, e.g., hard work causes fitness and *vice versa*, but again not in the same way, but the one as end, the other as the principle of motion."

104. *Physics* l. 1 t. 4 (184a23–26 in l. 1 c. 1).

105. Averroës, long commentary on *Physics* l. 1 t. 1 (184a10–16 in l. 1 c. 1), *AOAC v. 4 fol. 4 sg.* H—*fol. 6 sg.* H.

106. *Physics* l. 1 t. 1 (184a10–16 in l. 1 c. 1).

107. Zabarella's commentary on the *Physics*, *De naturalis scientiae constitutione*, was not published until 1586, though there, too, (*fol. 2, sg.* D) he refers to earlier remarks. Maybe the reference is to something he said in his 1568 lecture materials on *Physics* 8 (see note to *Regr. c. 6 par.* 5), but his comment at *Meth. l. 4 c. 20 par.* 9 suggests the existence of another body of commentary material that had been made public before 1578. See also the note to *Regr. c. 6 par.* 5.

108. *Nicomachean Ethics* l. 1 c. 13 (1102a5–1103a10 = l. 1 c. 13).

109. *Posterior Analytics* l. 1 t. 7–17 (71b9–72b4 = l. 1 c. 2).

110. Ibid. l. 1 t. 28–37 (73a21–74a4 = l. 1 c. 4).

111. Ibid. l. 1 t. 9 (71b19–25 in l. 1 c. 2).

112. Ibid. l. 2 t. 48–52 (94a20–b26 in l. 2 c. 11).

113. For "conjunction," compare *Regr. c.* 4 par. 9 and c. 9 par. 6–8.

114. Presumably the jurist Bartolus de Saxoferrato (1313/14–1357).

115. Those who seek a problem where there is none.

Bibliography

❧❧❧

EDITIONS OF *ON METHODS* AND *ON REGRESSUS*[*]

Opera logica. Venice: Paulo Meietti, 1578.

Opera logica. Editio Secunda. Venice: Giorgio Angelieri for Paulo Meietti, 1586.

Opera logica. In *Opera quae in hunc diem edidit*. Preface by Giulio Pace. [Frankfurt]: Jean Wechel for Jean Mareschal, 1586/87.

Opera logica. Preface by Johann Ludwig Hawenreuter. Basel: Konrad Waldkirch for Lazarus Zetzner and Pierre [and/or Jean] Mareschal, 1594.

Opera logica. Editio tertia. Preface by Johann Ludwig Hawenreuter. Cologne [but probably Neustadt, Basel, or Frankfurt]: Lazarus Zetzner, 1597. Photoreprint. Hildesheim: Olms, 1966.

Opera logica. Quarta editio. 3 vols. Venice: Paulo Meietti (vols. 1, 3) and Giovanni Antonio, Giacomo Franceschi, and Francesco Bolzetta (vol. 2), 1599–1601.

Opera logica. Editio Quarta. Preface by Johann Ludwig Hawenreuter. Cologne [but probably Neustadt, Basel, or Frankfurt]: Lazarus Zetzner 1602–3.

Opera logica. Venice: Paulo Meietti, and Treviso: Roberto Meietti, 1604.

Opera logica. Editio postrema. Preface by Johann Ludwig Hawenreuter. Frankfurt: Lazarus Zetzner, 1608.

Opera logica. Editio postrema. Preface by Johann Ludwig Hawenreuter. Frankfurt: Heirs of Lazarus Zetzner, 1622–23.

De methodis libri quatuor — Liber De regressu. With an introduction by Cesare Vasoli. Bologna: CLUEB, 1985. Contains a photoreprint of the 1578 edition.

---

[*] For an edition of the *Opera logica* printed in Venice in 1580, see Maclean, "Mediations of Zabarella," 56. A surviving copy of this imprint could not be located in a modern collection; it may be a ghost.

317

TRANSLATION

Schicker, Rudolf, trans. *Jacopo Zabarella: Über die Methoden (De methodis). Über den Rückgang (De regressu).* With introduction and notes by Schicker. Munich: Wilhelm Fink, 1995. Translation based on the 1578 edition, compared with the Frankfurt edition of 1608.

SECONDARY LITERATURE

Di Liscia, Daniel A., Eckhard Kessler, and Charlotte Methuen, eds. *Method and Order in Renaissance Philosophy of Nature: The Aristotle Commentary Tradition.* Aldershot: Ashgate, 1997.

Edwards, William F. "The Logic of Iacopo Zabarella (1533–1589)." PhD thesis, Columbia University, 1960.

Gilbert, Neal Ward. *Renaissance Concepts of Method.* New York: Columbia University Press, 1960.

Jardine, Nicholas. "Epistemology of the Sciences." In *The Cambridge History of Renaissance Philosophy*, 685–711, edited by Charles B. Schmitt, Quentin Skinner, Eckhard Kessler, and Jill Kraye. Cambridge: Cambridge University Press, 1988.

——. "Keeping Order in the School of Padua: Jacopo Zabarella and Francesco Piccolomini on the Offices of Philosophy." In Di Liscia et al., *Method and Order*, 183–210.

Maclean, Ian. "Mediations of Zabarella in Northern Europe, 1586–1623." Chap. 3 in *Learning and the Market Place.* Leiden: E. J. Brill, 2009.

Mikkeli, Heikki. *An Aristotelian Response to Renaissance Humanism. Jacopo Zabarella on the Nature of Arts and Sciences.* Helsinki: SHS, 1992. The only English-language monograph on Zabarella.

——. "Giacomo Zabarella." *The Stanford Encyclopedia of Philosophy* (Winter 2012 Edition). Edited by Edward N. Zalta. http://plato.stanford. edu/archives/win2012/entries/zabarella/.

Olivieri, Luigi, ed. *Aristotelismo veneto e scienza moderna: Atti del 250 anno academico del Centro per la storia della tradizione aristotelica nel Veneto.* 2 vols. Padua: Antenore, 1983.

Palmieri, Paolo. "Science and Authority in Giacomo Zabarella." *History of Science* 45 (2007): 404–27.

Piaia, Gregorio, ed. *La presenza dell'aristotelismo padovano nella filosofia della prima modernità: Atti del colloquio internazionale in memoria di Charles B. Schmitt*. Rome: Antenore, 2002.

Poppi, Antonino. *La dottrina della scienza in Giacomo Zabarella*. Padua: Antenore, 1972.

———. "Zabarella, or Aristotelianism as a Rigorous Science." In *The Impact of Aristotelianism on Modern Philosophy*, edited by Riccardo Pozzo, 35–63. Washington, D.C.: Catholic University of America Press, 2004.

Randall, John Herman. "The Development of Scientific Method in the School of Padua." *Journal of the History of Ideas* 1.2 (1940): 177–206. Reprinted in *The School of Padua and the Emergence of Modern Science*. Padua: Antenore, 1961.

Wallace, William A. "Circularity and the Paduan Regressus: From Pietro d'Abano to Galileo." *Vivarium* 33 (1995): 76–97.

———. "Randall redivivus: Galileo and the Paduan Aristotelians." *Journal of the History of Ideas* 49 (1988): 133–49.

Wear, Andrew, Roger Kenneth French, and Iain M. Lonie, eds. *The Medical Renaissance of the Sixteenth Century*. Cambridge: Cambridge University Press, 1985.

Index

꽃집꽃

Publication of this volume has been made possible by

The Myron and Sheila Gilmore Publication Fund at I Tatti
The Robert Lehman Endowment Fund
The Jean-François Malle Scholarly Programs and Publications Fund
The Andrew W. Mellon Scholarly Publications Fund
The Craig and Barbara Smyth Fund
for Scholarly Programs and Publications
The Lila Wallace–Reader's Digest Endowment Fund
The Malcolm Wiener Fund for Scholarly Programs and Publications